Inlets for Supersonic Missiles

John J. Mahoney

EDUCATION SERIES
J. S. Przemieniecki
Series Editor-in-Chief
Air Force Institute of Technology
Wright-Patterson Air Force Base, Ohio

Published by
American Institute of Aeronautics and Astronautics, Inc.
370 L'Enfant Promenade, SW, Washington, DC 20024

Library of Congress Cataloging-in-Publication Data

Mahoney, John J.
Inlets for supersonic missiles/John J. Mahoney.

p. cm.—(AIAA education series)
Includes bibliographical references and index.
1. Rockets (Aeronautics)—Ramjet engines—Air intakes. I. Title.
II. Series.
TL709.3R3M34 1990 623.4′519—dc20 90-45379
ISBN 0-930403-79-7

Texts in the AIAA Education Series

Re-Entry Vehicle Dynamics
  Frank J. Regan, 1984
Aerothermodynamics of Gas Turbine and Rocket Propulsion
  Gordon C. Oates, 1984
Aerothermodynamics of Aircraft Engine Components
  Gordon C. Oates, Editor, 1985
Fundamentals of Aircraft Combat Survivability Analysis and Design
  Robert E. Ball, 1985
Intake Aerodynamics
  J. Seddon and E. L. Goldsmith, 1985
Composite Materials for Aircraft Structures
  Brian C. Hoskins and Alan A. Baker, Editors, 1986
Gasdynamics: Theory and Applications
  George Emanuel, 1986
Aircraft Engine Design
  Jack D. Mattingly, William Heiser, and Daniel H. Daley, 1987
An Introduction to the Mathematics and Methods of Astrodynamics
  Richard H. Battin, 1987
Radar Electronic Warfare
  August Golden Jr., 1988
Advanced Classical Thermodynamics
  George Emanuel, 1988
Aerothermodynamics of Gas Turbine and Rocket Propulsion,
  Revised and Enlarged
  Gordon C. Oates, 1988
Re-Entry Aerodynamics
  Wilbur L. Hankey, 1988
Mechanical Reliability: Theory, Models and Applications
  B. S. Dhillon, 1988
Aircraft Landing Gear Design: Principles and Practices
  Norman S. Currey, 1988
Gust Loads on Aircraft: Concepts and Applications
  Frederic M. Hoblit, 1988
Aircraft Design: A Conceptual Approach
  Daniel P. Raymer, 1989
Boundary Layers
  A. D. Young, 1989
Aircraft Propulsion Systems Technology and Design
  Gordon C. Oates, Editor, 1989
Basic Helicopter Aerodynamics
  J. Seddon, 1990
Introduction to Mathematical Methods in Defense Analyses
  J. S. Przemieniecki, 1990
Space Vehicle Design
  Michael D. Griffin and James R. French, 1991
Inlets for Supersonic Missiles
  John J. Mahoney, 1991

Published by
American Institute of Aeronautics and Astronautics, Inc., Washington, DC

FOREWORD

Inlets for Supersonic Missiles by John J. Mahoney complements the text *Intake Aerodynamics* by J. Seddon and E. L. Goldsmith published previously in the AIAA Education Series. While the earlier text covered the broad spectrum of speed ranges from low subsonic to high supersonic flight regimes, Mahoney concentrates on the application of inlet technology to supersonic missiles that are air-launched at high-subsonic Mach number and rocket-boosted to supersonic velocity, where a ramjet engine sustains flight in the atmosphere in the Mach number range between 2 and 4. He does not cover inlet design for supersonic-combustion ramjet (Scramjet) technology; but that subject will be treated in a coming Education Series text on hypersonic propulsion.

Mahoney's text reflects his long experience with the design and development of missile inlets. He gives a detailed description of the inlet selection and design process, covering such design variables as number of inlets, placement of the missile on the launch aircraft, design Mach number, and mission profile. Providing a wealth of information for designers and research engineers alike, Mahoney's book should be useful both as a basic course text and as a reference work on the design of supersonic inlets.

J. S. PRZEMIENIECKI
Editor-in-Chief
AIAA Education Series

TABLE OF CONTENTS

PREFACE

This book describes the design, operation, performance, and selection of the inlets (also known as intakes and air-induction systems) indispensable to proper functioning of an air-breathing engine. It treats in detail interrelationships between an inlet and other engine components and interactions between an installed inlet and vehicle components, because these factors constitute an integral part of the inlet-development process. Although the material presented has timeless relevance to inlets in general, the main theme and supporting examples focus on the development of fixed-geometry members of the axisymmetric, two-dimensional, and three-dimensional families. These inlets principally see application to missiles that are air-launched at high-subsonic Mach number and rocket-boosted to supersonic velocity where a ramjet engine sustains flight in the Mach-number range of 2–4. But the text also explains and explores ramifications of extension of missile operations into the hypersonic regime by the use of variable-geometry inlets and exit nozzles. All ramjet engines described here operate with subsonic flow through the combustor component. The supersonic combustion ramjet (Scramjet) has different requirements than those addressed in this book.

The aerodynamic-design procedures, operational characteristics, and performance-estimating techniques described incorporate proven analytical methods, although much information has heretofore been resident in well-worn notebooks—i.e., rules of thumb and shortcuts that have not yet been formalized or derived in rigorous mathematical terms, and experimental data that were unreported or were contained in reports that were classified at the time of publication. (All of the material presented in this book is now unclassified.) The discussion of the inlet-selection process includes such variables as number of inlets, family, placement on vehicle, design Mach number, supersonic compression direction, and influence of the launch aircraft and the missile mission.

The author is indebted to L. C. Dunsworth, who has retired from The Marquardt Co., for technical editing and constant encouragement; and to A. E. Heins, also retired from Marquardt, for contributing the three-dimensional inlet design procedures. Much help in locating reports, reference information, and technical details was provided by colleagues in such diverse places as Marquardt, United Technologies Research Center, Chemical Systems Division of United Technologies, McDonnell Douglas Astronautics Company-St. Louis, The Boeing Company, The Vought

Company, LTV Corporation, the U.S. Naval Weapons Center, the U.S. Naval Surface Weapons Center, The Applied Physics Laboratory of Johns Hopkins University, the Air Force Aeropropulsion Laboratory, and the Ames, Lewis, and Langley Research Centers of NASA. The author also thanks Marge Old, who typed the manuscript, and Antonella Thompson, who prepared the final figures.

JOHN J. MAHONEY
JANUARY 1991

1
FUNCTIONS AND FUNDAMENTALS

1.1 Introduction

Most manned aircraft and many missiles rely on some form of air-breathing propulsion for sustained flight within the Earth's atmosphere. All air-breathing engines require an inlet to diffuse air from freestream velocity to a lower velocity acceptable for further processing by other engine components. The inlet component is designed to capture the exact amount of air required, accomplish the diffusion with minimum total pressure loss, deliver the air with tolerable flow distortion, and contribute the least possible external drag to the system.

The basic purpose of all air-breathing engines is to convert fuel energy into useful work done on an airplane or missile. The most popular modern form of an air-breathing engine utilizes jet propulsion wherein the captured freestream air mass is diffused in the inlet, mixed with fuel and burned in a combustion chamber, and expanded through a nozzle back to ambient pressure. The in-line arrangement of an inlet, combustor, and nozzle to form a typical flow-through ramjet engine is shown in Fig. 1.1. With the installation of a compressor between the inlet and combustor components and a turbine between the combustor and nozzle, a turbojet engine is formed as shown in Fig. 1.2.

1.2 Brayton Cycle

Both the ramjet and turbojet operate on the Brayton cycle, which is shown in Fig. 1.3 in schematic form on an enthalpy-entropy (H-S) diagram for the ramjet engine. Lines of constant pressure are identified as P_a, P_b, and P_c. Irreversible adiabatic compression occurs in the inlet between points 0 and 2; nearly constant pressure combustion takes place in the combustion chamber between points 2 and 4; and isentropic expansion to ambient pressure is achieved across the nozzle from points 4–6. Point 0 represents the condition of ambient air at rest. The kinetic energy of the air captured by the inlet is represented by the term $V_{\mathrm{INLET}}^2/2g$, and the magnitude is determined by the velocity of the vehicle relative to the ambient air. During the compression, or diffusion, process, the inlet converts kinetic energy to potential energy, and, in accordance with the law

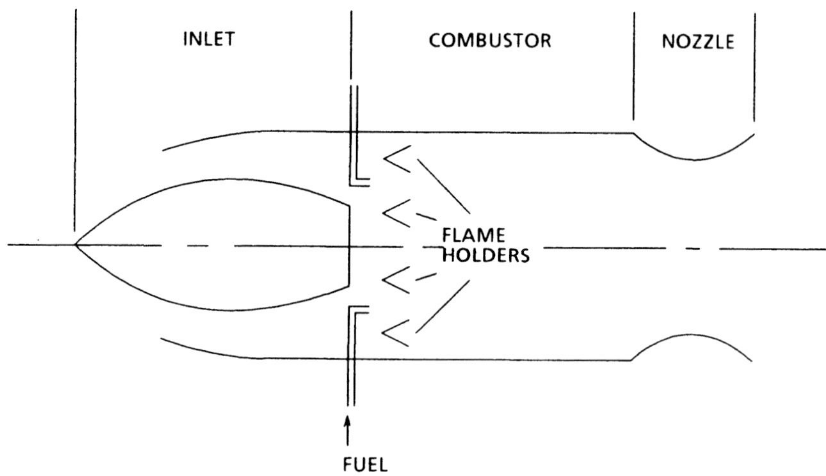

Fig. 1.1 Typical ramjet engine sketch.

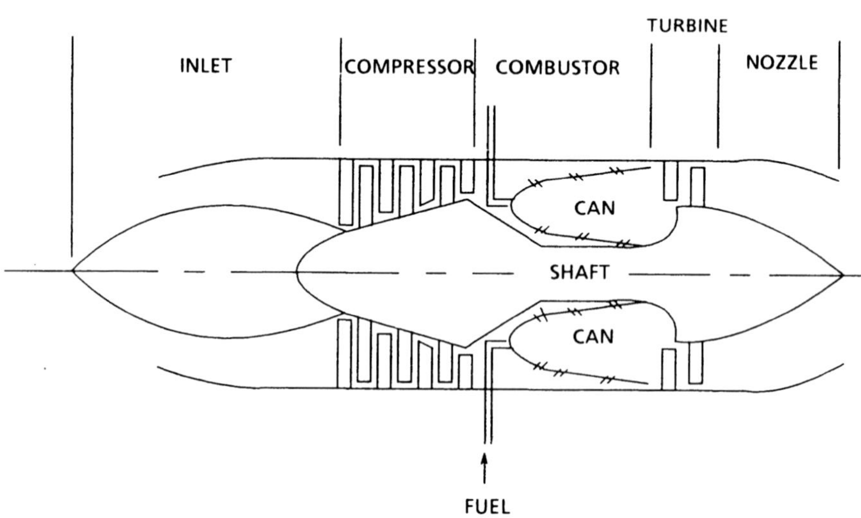

Fig. 1.2 Typical turbojet engine sketch.

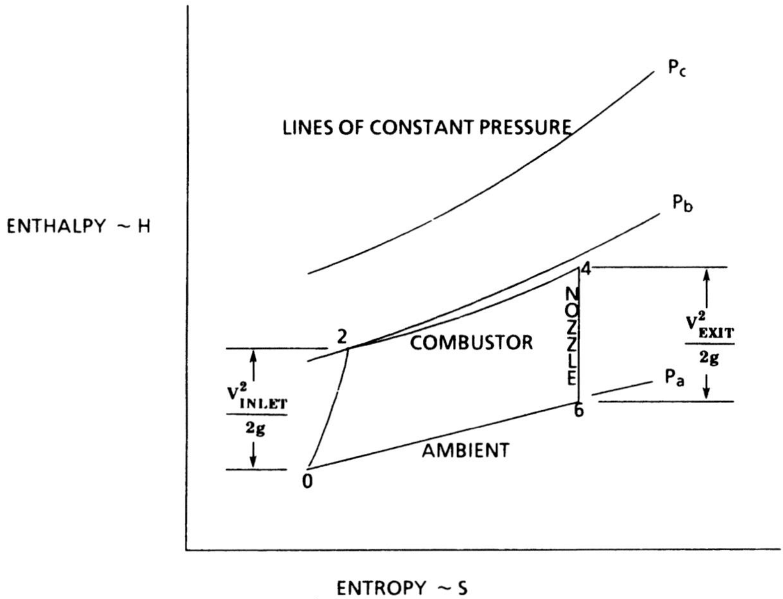

Fig. 1.3 Schematic of Brayton cycle for a typical ramjet engine.

of conservation of energy, the static enthalpy h plus the kinetic energy equals the total enthalpy H at points 0 and 2:

$$h_0 + \frac{V^2_{\text{INLET}}}{2g} = H_0 = H_2 \tag{1.1}$$

which, for a thermally perfect gas, corresponds to

$$C_P Tt_0 = C_P Tt_2 \tag{1.2}$$

where C_P is the specific heat at constant pressure, and Tt the total temperature.

Because of the dissipation of mechanical energy into heat through viscous effects in any inlet, there is an increase in entropy between points 0 and 2, and the stagnation pressure at point 2 is always less than that at point 0:

$$Pt_2 < Pt_0 \tag{1.3}$$

Pressure loss in the combustor component is reflected by the lower pressure at point 4 relative to point 2. Pressure loss also contributes to the entropy increase between points 2 and 4, although the bulk of the entropy gain is caused by heat addition. The combustion process raises the enthalpy level of the cycle from H_2 to H_4, and the nozzle converts potential energy to kinetic energy so that

$$h_6 + \frac{V_{\mathrm{EXIT}}^2}{2g} = H_6 = H_4 \tag{1.4}$$

or

$$C_P Tt_6 = C_P Tt_4 \tag{1.5}$$

Because the expansion process is very nearly isentropic, there is no significant entropy increase, and

$$Pt_6 = Pt_4 \tag{1.6}$$

The difference between the magnitudes of the kinetic energy of the exiting gas, $V_{\mathrm{EXIT}}^2/2g$, and the kinetic energy of the captured air mass, $V_{\mathrm{INLET}}^2/2g$, represents the net energy available to do work.

If the magnitude of the term $V_{\mathrm{INLET}}^2/2g$ should decrease to zero (i.e., a vehicle at rest), then the ramjet can produce no thrust. At zero forward velocity, there can be no mass flow through the engine, and the internal engine pressure will be no greater than ambient. A compressor installed between the inlet and combustor components can increase engine pressure above the ambient level even at zero forward velocity and raise the enthalpy level of the cycle such that the combustion process can be sustained. The compressor is rotated by a turbine mounted on the same shaft and installed between the combustor and nozzle components, as shown in Fig. 1.2. Heat energy is converted into mechanical energy to turn the shaft by means of expansion of the products of combustion through the turbine wheel buckets. At forward speeds above zero, the compressor augments the compression achieved by the inlet.

1.3 Thrust Determination

The net energy available to do work as determined from inspection of the Brayton cycle is related to thrust by means of the expression for propulsive efficiency

$$n_P = \frac{\text{Useful work done on vehicle}}{\text{Energy available to do work}} \tag{1.7}$$

or

$$n_P = \frac{FV_0}{\dfrac{\dot{w}(V_6^2 - V_0^2)}{2g}} \tag{1.8}$$

where F is the thrust (lbs), V_0 the velocity of the vehicle (ft/s), $\dot{w}$ the weight flow of air (lbs/s), and V_6 the exit velocity of the products of combustion (ft/s).

Engine thrust is produced in accordance with Newton's law wherein the thrust force is equal to the rate of change of momentum of the air (plus fuel) mass passing through the engine. As the first step in the development of the mathematical relationship to be used herein, it is appropriate to identify key stations in a typical jet engine. A flow-through ramjet has been chosen as the representative engine and is shown in the schematic in Fig. 1.4. Station 0 represents freestream, or ambient, conditions; station 1 or C is the inlet cowl lip station; station 2 is the end of the inlet component; station 4 is the end of the combustor; station 5 is the exit nozzle throat; and station 6 is the end of the exit nozzle. Note that stations 0, 2, 4, and 6 correspond to the key points in the Brayton cycle diagram. Station 3 represents the end of the fuel injector/flame holding section of the combustor and is used as a bookkeeping station to define the arbitrary boundary between aerodynamic pressure losses and heat addition pressure losses in the combustor component.

Engine thrust force F is defined as the exit stream thrust (momentum term plus pressure area term) at station 6 minus the entering stream thrust at station 0 minus the integral of the external drag forces acting on all of the engine components.

Mathematically,

$$F = \rho_6 A_6 V_6^2 + P_6 A_6 - (\rho_0 A_0 V_0^2 + P_0 A_0) - \int_0^6 [(P_A \sin\theta) + \tau_A]\, dA_w \tag{1.9}$$

where

ρ = mass per cubic foot (lb-s²/ft⁴)
A = area (ft²)
V = velocity (ft/s)
P = unit pressure (lb/ft²)
P_A = unit pressure drag (lb/ft²)
τ_A = friction drag (lb/ft²)
dA_w = element of wetted surface area (ft²)
θ = local surface angle (deg)

and the two latter symbols are defined graphically in Fig. 1.4.

Equation (1.9) is valid for the turbojet as well as the ramjet. A more convenient expression for practical use can be developed as follows:

The equation of state relates gas properties:

$$\rho = \frac{P}{RT} \tag{1.10}$$

where R is the gas constant ($= 1716.322\ \mathrm{ft^2/s^2-°R}$), and T the temperature (°R).

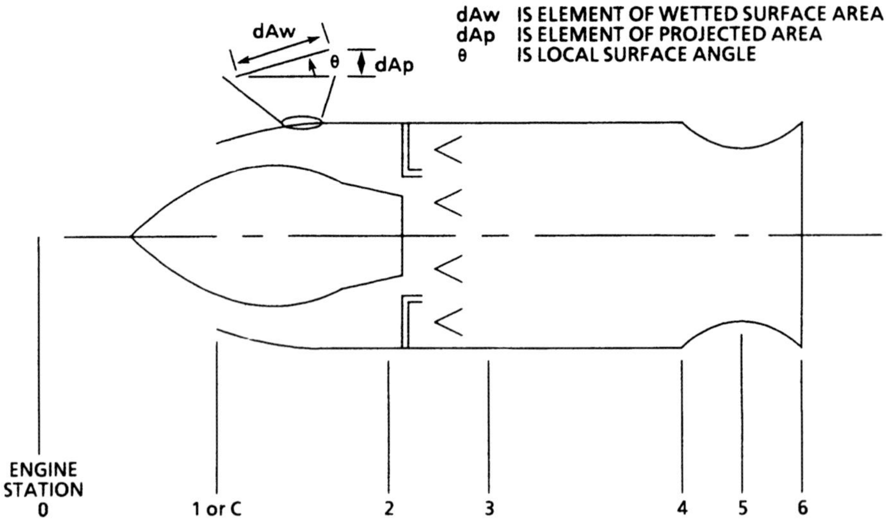

Fig. 1.4 Representative ramjet engine schematic.

An index of the compressibility of the flow is determined from

$$V = aM \tag{1.11}$$

where a is the velocity of sound (ft/s), and M the Mach number. The velocity of sound varies in accordance with

$$a = \sqrt{\gamma RT} \tag{1.12}$$

where γ is the ratio of specific heats C_P/C_V and is 1.4 for air near 0°F.

Substituting Eqs. (1.10), (1.11), and (1.12) into Eq. (1.9), ignoring the drag term for the present, and referencing the remaining terms to ambient pressure P_0 yields

$$F = \frac{P_6}{RT_6} A_6 \gamma_6 RT_6 M_6^2 + P_6 A_6 - P_0 A_6 - \frac{P_0}{RT_0} A_0 \gamma_0 RT_0 M_0^2$$

Clearing and combining terms,

$$F = P_6 A_6 (1 + \gamma_6 M_6^2) - P_0 A_0 \gamma_0 M_0^2 - P_0 A_6 \tag{1.13}$$

The expression can be nondimensionalized by specifying a thrust coefficient similar to the coefficient forms used for the other body forces (lift and drag):

$$C_F = \frac{F}{q_0 A_R} \tag{1.14}$$

where q_0 is the dynamic pressure,

$$= \tfrac{1}{2}\rho_0 V_0^2$$

$$= \frac{\gamma_0}{2} P_0 M_0^2 \qquad (\text{lb/ft}^2)$$

and A_R is an arbitrary reference area (ft^2).

Substituting Eq. (1.13) into Eq. (1.14),

$$C_F = \frac{2(A_6/A_R)}{\gamma_0 M_0^2}\left[\frac{P_6}{P_0}(1 + \gamma_6 M_6^2) - 1\right] - 2\frac{A_0}{A_R} \tag{1.15}$$

But, P_6/P_0 may be written as

$$\frac{\dfrac{Pt_2}{Pt_0}\dfrac{Pt_4}{Pt_2}\dfrac{Pt_6}{Pt_4}\dfrac{P_6}{Pt_6}}{\dfrac{P_0}{Pt_0}}$$

and

$$\frac{A_0}{A_R} = \frac{A_0}{A_C}\frac{A_C}{A_R}$$

Thus,

$$C_F = \frac{2(A_6/A_R)}{\gamma_0 M_0^2}\left[\frac{\dfrac{Pt_2}{Pt_0}\dfrac{Pt_4}{Pt_2}\dfrac{Pt_6}{Pt_4}\dfrac{P_6}{Pt_6}(1 + \gamma_6 M_6^2)}{\dfrac{P_0}{Pt_0}} - 1\right] - 2\frac{A_0}{A_C}\frac{A_C}{A_R}$$

Defining a nozzle efficiency,

$$n_N = \frac{\dfrac{P_6}{Pt_4}(1 + \gamma_6 M_6^2)_{\text{actual}}}{\dfrac{P_6}{Pt_6}(1 + \gamma_6 M_6^2)_{\text{ideal}}} \tag{1.16}$$

The denominator is tabulated as stream thrust function, f/Pt, in Ref. 2.

Then,

$$C_F = \frac{2(A_6/A_R)}{\gamma_0 M_0^2}\left\{\frac{\dfrac{Pt_2}{Pt_0}\dfrac{Pt_4}{Pt_2}n_N\left[\dfrac{P_6}{Pt_6}(1 + \gamma_6 M_6^2)_{\text{ideal}}\right]}{\dfrac{P_0}{Pt_0}} - 1\right\} - 2\frac{A_0}{A_C}\frac{A_C}{A_R} \tag{1.17}$$

Thus, the thrust of a jet engine can be expressed nondimensionally in coefficient form in terms of area ratios, pressure ratios, and Mach number functions. The contribution of each major engine component may be determined by examining the individual terms in Eq. (1.17).

The selection of a freestream Mach number M_0 determines the value of the term M_0^2 and the term P_0/Pt_0, which is related to M_0 through the Bernoulli equation for compressible fluid flow:

$$\frac{P_0}{Pt_0} = \left[\frac{\gamma - 1}{2}M_0^2 + 1\right]^{-[\gamma/(\gamma-1)]} \tag{1.18}$$

Tabulated solutions to Eq. (1.18) are given in Ref. 1 over a range of Mach numbers from 0 to 100.

Another independent variable is the nozzle exit area A_6. If all other terms remain constant, an increase in A_6 relative to the reference area A_R produces a proportionate increase in C_F. In practice, however, increases in A_6 above the cross-sectional area of the combustor (A_3 in the schematic of Fig. 1.4) increase the external surface angle and the concomitant external drag, which offsets the gains in internal thrust.

The second independent nozzle variable is throat size A_5, which, for a fixed-geometry nozzle, is usually selected to yield maximum engine thrust at the freestream design Mach number of the inlet. For modern integral engines a typical average value of A_5 is 0.50 referenced to the cross-sectional area A_3 of the combustor. The nozzle expansion area ratio A_6/A_5 uniquely defines the ideal exit velocity $V_{6_{id}}$ and Mach number $M_{6_{id}}$, which establishes the value of the exit stream thrust term, $P_6/Pt_6\ (1 + \gamma_6 M_6^2)$ ideal. The use of an ideal stream thrust term is consistent with the isentropic expansion shown on the Brayton cycle diagram between points 4 and 6. Deviations from the isentropic (viscous losses and divergence losses in conical nozzles) are lumped in the nozzle efficiency term n_N, which is usually assigned a value of 0.98 based on experimental results with a variety of convergent/divergent nozzles.

The last independent variable in the equation is the cross-sectional area A_0 of the freestream tube captured by the inlet. To increase the thrust coefficient, it is obviously desirable to minimize A_0, but decreases in A_0 are usually accompanied by increases in external drag that offset the thrust gain. Also, A_0 must be sized in consonance with A_5 to achieve inlet critical pressure recovery at the design condition. The inlet also contributes to engine thrust through the pressure recovery term Pt_2/Pt_0, and a high pressure recovery is an index of a "good" inlet. Similarly, a high value of combustor pressure recovery, Pt_4/Pt_2, is desirable for maximum thrust. These two pressure recovery terms are dependent variables, however, and are unknown quantities in Eq. (1.17). Obviously, another equation is necessary to determine values.

1.4 Continuity Equation

The cross-sectional area A_0 of the freestream tube captured by the inlet, the nozzle throat size A_5, the pressure recovery across the engine, which is the product of Pt_2/Pt_0 and Pt_4/Pt_2, the fuel mass injected into the engine, and the heat energy added to the cycle in the combustion chamber are interrelated by the law of conservation of mass. Simply stated, the mass flow crossing any station in the engine must equal the mass flow crossing any other station with appropriate accounting for mass either added or subtracted between the two stations. Inspection of the ramjet engine schematic in Fig. 1.4 indicates that the two most logical stations in terms of known flow properties are freestream station 0 and nozzle throat station 5.

At station 0, where the flow is uniform and parallel, the air weight flow rate is defined as

$$\dot{w}_0 = \rho_0 A_0 V_0 \tag{1.19}$$

Substituting into Eq. (1.19) the relationships previously defined in Eqs. (1.10), (1.11), and (1.12) yields

$$\dot{w}_0 = g\sqrt{\frac{\gamma}{R}}\frac{P_0}{\sqrt{T_0}}A_0 M_0$$

If the numerator and denominator are multiplied by Pt_0 and $\sqrt{(Tt_0)}$, the result is

$$\dot{w}_0 = Pt_0\frac{P_0}{Pt_0}g\sqrt{\frac{\gamma}{R}}M_0\sqrt{\frac{Tt_0}{T_0}}\frac{A_0}{\sqrt{Tt_0}}$$

A form of the energy equation is

$$\sqrt{\frac{Tt_0}{T_0}} = \left(1 + \frac{\gamma - 1}{2}M_0^2\right)^{1/2} \tag{1.20}$$

Then,

$$\dot{w}_0 = Pt_0\frac{P_0}{Pt_0}\left[g\sqrt{\frac{\gamma}{R}}M_0\left(1 + \frac{\gamma - 1}{2}M_0^2\right)^{1/2}\right]\frac{A_0}{\sqrt{Tt_0}}$$

The term in the brackets has been assigned the symbol $\mathring{m}$ and is designated mass flow function, which is tabulated over a range of Mach numbers from 0 to 5.0 in Ref. 2. Also tabulated in the same reference is the product of $[(P/Pt)\mathring{m}]$. Thus, Eq. (1.19) may be written as

$$\dot{w}_0 = \frac{Pt_0[(P/Pt)\mathring{m}]_0 A_0}{\sqrt{Tt_0}} \tag{1.21}$$

At station 5, the pressure ratio across the nozzle is normally high enough to ensure choked flow, so that the Mach number for the ideal nozzle is unity. The product of combustion weight flow is defined as

$$\dot{w}_5 = \rho_5 A_5 V_5 C_{N_M} \tag{1.22}$$

where C_{N_M} is the nozzle mass flow coefficient, which equals the actual mass flow through nozzle throat divided by the theoretical mass flow through the geometric nozzle throat, $\cong 0.97$.

Substituting into Eq. (1.22), the same relationships used previously in the station 0 air weight flow equation yield

$$\dot{w}_5 = \frac{Pt_5[(P/Pt)\mathring{m}]_5 A_5 C_{N_M}}{\sqrt{Tt_5}} \tag{1.23}$$

where A_5 is the geometric nozzle throat, which passes a uniform, parallel flow at $M_5 = 1.0$. Distortions of the throat flow profile caused by viscous effects decrease the flow rate below the ideal value, and the C_{N_M} is a measure of the difference.

From continuity,

$$\dot{w}_5 = \dot{w}_0 + f/a\dot{w}_0 \tag{1.24}$$

where f/a is the fuel-to-air ratio.

Substituting Eqs. (1.21) and (1.23) into Eq. (1.24),

$$\frac{Pt_5\left(\dfrac{P}{Pt}\mathring{m}\right)_5 A_5 C_{N_M}}{\sqrt{Tt_5}} = \left(1 + \frac{f}{a}\right)\frac{Pt_0\left(\dfrac{P}{Pt}\mathring{m}\right)_0 A_0}{\sqrt{Tt_0}} \tag{1.25}$$

If the engine geometry has been selected, i.e., A_5 and A_0 are known, Eq. (1.25) can be solved for the pressure recovery across the engine:

$$\frac{Pt_5}{Pt_0} = \frac{\left(\dfrac{P}{Pt}\mathring{m}\right)_0 A_0(1 + f/a)}{\left(\dfrac{P}{Pt}\mathring{m}\right)_5 A_5 C_{N_M}}\sqrt{\frac{Tt_5}{Tt_0}} \tag{1.26}$$

where

$$\frac{Pt_5}{Pt_0} = \frac{Pt_5}{Pt_4}\frac{Pt_4}{Pt_2}\frac{Pt_2}{Pt_0}$$

but

$$\frac{Pt_5}{Pt_4} = 1.0 \qquad \text{[Eq. (1.6)]}$$

and

$$\frac{Pt_4}{Pt_2}\frac{Pt_2}{Pt_0}$$

is the product of the combustor and inlet pressure recoveries, respectively, as shown in the thrust coefficient equation [Eq. (1.17)].

$$\left(\frac{P}{Pt}\mathring{m}\right)_0$$

is a function of only the selected freestream Mach number M_0, and

$$\left(\frac{P}{Pt}\mathring{m}\right)_5$$

is uniquely defined by $M_5 = 1.0$, and the value is given in the "hot deck" ($\gamma = 9/7$) of Ref. 2. The variables A_0 and $A_5C_{N_M}$ are known, f/a is an independent variable, and

$$\sqrt{\frac{Tt_5}{Tt_0}}$$

is a function of f/a as developed later in this chapter.

A review of the Brayton cycle diagram of Fig. 1.3 shows that the law of conservation of energy requires that $C_PTt_0 = C_PTt_2$ [Eq. (1.2)] and that $C_PTt_4 = C_PTt_5 = C_PTt_6$. In essence, the enthalpy increase across the combustor can be represented by a total temperature difference, $Tt_4 - Tt_2$, or by $Tt_5 - Tt_0$, which may also be written simply as ΔT, and

$$Tt_5 = Tt_0 + \Delta T \tag{1.27}$$

but

$$Tt_0 = T_0\left(1 + \frac{\gamma - 1}{2}M_0^2\right)$$

where T_0 is the ambient temperature at any selected altitude, and M_0 is any selected freestream Mach number. Values of T/Tt as a function of M are tabulated in Refs. 1 and 2.

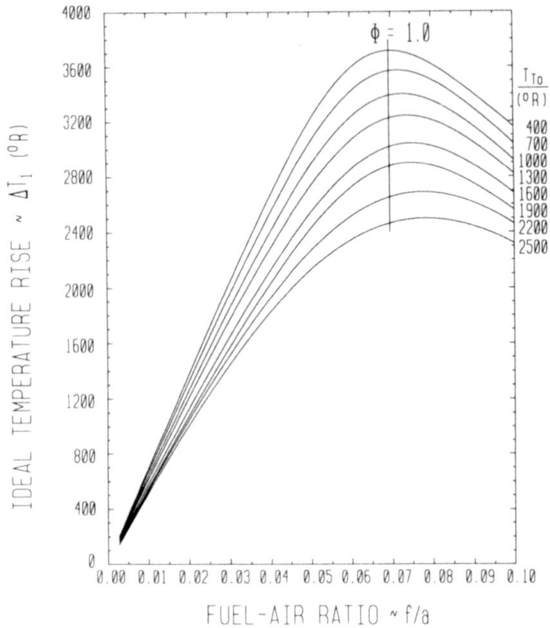

Fig. 1.5 Ideal temperature rise curves for JP-5 fuel.

For any fuel it is possible to construct ideal temperature rise curves based on equilibrium thermochemistry as a function of the fuel-to-air ratio for constant pressure combustion with air at any initial (freestream) stagnation temperature Tt_0. Figure 1.5 illustrates typical variations of ideal temperature rise curves for JP-5. The actual temperature rise achieved in any combustor is always less than the ideal value, and the ratio of the actual to ideal defines the combustion efficiency:

$$n_c = \frac{\Delta T_{\text{actual}}}{\Delta T_{\text{ideal}}} = \frac{Tt_{5_{\text{actual}}} - Tt_2}{Tt_{5_{\text{ideal}}} - Tt_2} \tag{1.28}$$

Equation (1.27) becomes

$$Tt_5 = Tt_0 + n_c(\Delta T_i) \tag{1.29}$$

where n_c is a function of a host of variables including combustor geometry, fuel distribution, fuel penetration, flame holding and flame spreading, turbulence, fuel and air mixture, combustion severity, and altitude. Most of these variables are relatively intractable to analytical procedures; hence, the traditional approach has been to rely on empirical methods or experimental measurements with a specific combustor configuration operated over a prescribed range of test conditions.

For any given level of combustion efficiency n_c and freestream stagnation temperature Tt_0, the temperature at the completion of combustion Tt_5 varies with the ideal temperature rise ΔT_i, which increases with increasing f/a ratio from zero at no fuel combustion to a maximum value at the point labeled $\phi = 1$ on Fig. 1.5. Complete combustion is achieved when there is sufficient oxygen present to combine with all of the fuel, and this condition is called stoichiometric fuel-to-air ratio, or an equivalence ratio (ϕ) of unity. At ϕ above unity, there is an excess of fuel, which adds mass but no heat, and the ΔT curves begin to decrease in magnitude.

1.5 Engine Performance Summary

Engine performance can be completely described by three equations: the nondimensional thrust coefficient equation [Eq. (1.17)], the continuity equation [Eq. (1.26)], and an expression that relates the pounds of fuel consumed per hour to the pounds of thrust produced:

$$C_F = \frac{2(A_6/A_R)}{\gamma_0 M_0^2} \left\{ \frac{\dfrac{Pt_2}{Pt_0}\dfrac{Pt_4}{Pt_2} n_N \left[\dfrac{P_6}{Pt_6}(1 + \gamma_6 M_6^2)_{\text{ideal}} \right]}{\dfrac{P_0}{Pt_0}} - 1 \right\} - 2\frac{A_0}{A_C}\frac{A_C}{A_R} \tag{1.17}$$

$$\frac{Pt_5}{Pt_0} = \frac{\left(\dfrac{P}{Pt}\mathring{m}\right)_0 A_0(1 + f/a)}{\left(\dfrac{P}{Pt}\mathring{m}\right)_5 A_5 C_{N_M}} \sqrt{\frac{Tt_5}{Tt_0}} \tag{1.26}$$

where

$$\frac{Pt_5}{Pt_0} = \frac{Pt_2}{Pt_0}\frac{Pt_4}{Pt_2}$$

and the specific fuel consumption,

$$\text{SFC} = \frac{3600\dot{w}_f}{C_F q_0 A_R} \tag{1.30}$$

where $\dot{w}_f$ is the weight flow of fuel (lbs/s) $= f/a\dot{w}_0$.

It has been shown that increasing f/a up to the stoichiometric value will increase Tt_5 [Eq. (1.29)], and concomitantly the term in Eq. (1.26), $(1 + f/a)\sqrt{Tt_5/Tt_0}$, also called the heat addition parameter. Thus, for an engine for which freestream capture area and nozzle throat size have been selected, increasing fuel flow rate increases the pressure recovery across the engine and thereby increases engine thrust if the nozzle exit size has also been selected.

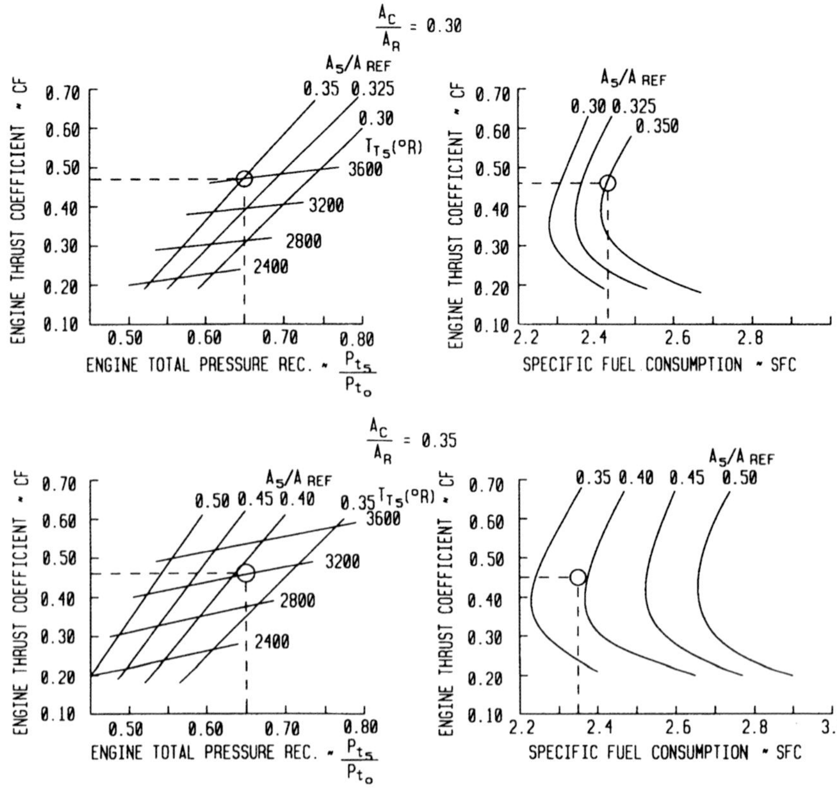

Fig. 1.6 Performance map for a ramjet engine operating at Mach 2.5 at sea level as a function of component size and fuel flow rate variations.

For any desired combination of M_0 and altitude (P_0, T_0), a map of engine performance (C_F, Pt_5/Pt_0, and SFC) can be developed for any suitable fuel at any selected fuel flow rate (expressed in terms of Tt_5) and with any reasonable engine size as determined by A_C, A_3, A_5, and A_6, all referenced to an arbitrarily selected area, A_R. Such a map is shown in Fig. 1.6 for a ramjet engine operating at Mach 2.5 at sea level. The left-hand plots on the map show the variation of C_F with Pt_5/Pt_0 for four different levels of Tt_5 and a variety of nozzle throat sizes, A_5/A_R. The right-hand plots show the relationship between C_F and SFC as a function of the same variables. The two upper plots are commensurate with an inlet cowl size, A_C/A_R, of 0.30, whereas the two lower plots are compatible with an $A_C/A_R = 0.35$. If the thrust coefficient required to maintain a steady-state cruise condition at Mach 2.5 and sea level has a value of 0.45, and the maximum allowable total pressure recovery across the engine is 0.65, then the smaller inlet in the upper plots requires a smaller nozzle throat size (as represented by the small circle at the intersection of the dashed lines) than the larger inlet. However, the smaller inlet requires the engine to operate with a higher fuel flow rate (higher Tt_5) and hence with a higher specific fuel consumption. Thus, the designer has a choice between the larger, more economical engine and the smaller, presumably lighter in weight engine with the potential for less drag.

The thrust coefficient defined in Eq. (1.17) is termed "net jet" and accounts only for internal thrust. Equation (1.9), which accounts for the external drag forces in the term

$$\int_0^6 [(P_A \sin\theta) + \tau_A] \, dA_w$$

is the expression for net thrust. The designer's choice would be influenced by the drag forces generated in each of the engine cases given earlier. The methods for estimating drag will be discussed in subsequent chapters.

Implicit in the development of the engine performance equations has been the concept of one-dimensional flow where parameters are assumed to vary extremely slowly in all directions except the axial. This concept has greatly simplified solutions to the nonlinear partial differential equations that define the flow through and around an engine. The simplified performance equations presented herein have been proven sufficiently accurate to satisfy all engineering requirements. The one-dimensional concept also provides a simple and practical approach to the analysis of the inner workings of the major engine components.

For example, the flow through a nozzle from a high pressure in the combustion chamber to ambient pressure at the exit plane can be exactly defined based on one-dimensional, isentropic expansion through a choked throat. From the continuity equation the mass flow crossing any nozzle station must equal the mass flow crossing the choked throat station:

$$\rho A V = \rho^* A^* V^* \tag{1.31}$$

where the asterisk defines conditions where $M = 1.0$.

Substituting Eqs. (1.10), (1.11), and (1.12) into (1.31),

$$\frac{\mathrm{PAM}}{\sqrt{T}} = \frac{P^* A^*}{\sqrt{T^*}} \tag{1.32}$$

Multiplying both sides by $\sqrt{(Tt)}/(Pt)$,

$$\frac{P}{Pt}\sqrt{\frac{Tt}{T}}\,AM = \frac{P^*}{Pt^*}\sqrt{\frac{Tt^*}{T^*}}\,A^* \tag{1.33}$$

[where $\sqrt{(Tt)} = \sqrt{(Tt)^*}$ from the conservation of energy and $Pt = Pt^*$ for isentropic flow], and substituting into Eq. (1.33) the relationships shown in Eqs. (1.18) and (1.20),

$$\frac{P}{Pt} = \left[\frac{\gamma - 1}{2}M^2 + 1\right]^{-[\gamma/(\gamma-1)]}$$

$$\frac{P^*}{Pt^*} = \left[\frac{\gamma - 1}{2} + 1\right]^{-[\gamma/(\gamma-1)]} = \left[\frac{\gamma + 1}{2}\right]^{-[\gamma/(\gamma-1)]}$$

$$\sqrt{\frac{Tt}{T}} = \left[1 + \frac{\gamma - 1}{2}M^2\right]^{1/2}$$

$$\sqrt{\frac{Tt^*}{T^*}} = \left[1 + \frac{\gamma - 1}{2}\right]^{1/2} = \left[\frac{\gamma + 1}{2}\right]^{1/2}$$

yields

$$\left[\frac{\gamma - 1}{2}M^2 + 1\right]^{-[\gamma/(\gamma-1)]}\left[1 + \frac{\gamma - 1}{2}M^2\right]^{1/2} AM = \left[\frac{\gamma + 1}{2}\right]^{-[\gamma/(\gamma-1)]}\left[\frac{\gamma + 1}{2}\right]^{1/2} A^* \tag{1.34}$$

Solving Eq. (1.34) for A/A^*,

$$\frac{A}{A^*} = \frac{1}{M}\left[\frac{2 + (\gamma - 1)\,M^2}{\gamma + 1}\right]^{(\gamma+1)/[2(\gamma-1)]} \tag{1.35}$$

At any station in the nozzle where the cross-sectional area is known relative to the throat area, Eq. (1.35) can be solved to determine the subsonic Mach number at any station upstream from the throat and the supersonic Mach number at any station downstream from the throat. Conversely, of course, the area can be determined from a known Mach number. Once the Mach number is known, all other flow properties (pressure, temperature, density, velocity, and speed of sound) are uniquely defined at any nozzle station.

The one-dimensional flow picture is also useful in analyzing the total pressure losses attributable to the combustor component. As shown in the thrust coefficient equation [Eq. (1.17)], a high value of combustor total pressure recovery, Pt_4/Pt_2, is conducive to a high thrust level. The overall combustor total pressure recovery can be arbitrarily divided into two terms that reflect the separate effects of aerodynamic pressure losses and heat addition pressure losses and that can be accounted for by means of the bookkeeping station 3 shown on the engine schematic of Fig. 1.4. Aerodynamic losses are caused by internal drag associated with such elements as wall blockage, fuel injector blockage, flame holder blockage, and fuel addition, and these losses are assumed to occur between stations 2 and 3 in Fig. 1.4. Heat addition losses are caused by an increase in entropy between stations 3 and 4. Thus, the overall combustor total pressure recovery, Pt_4/Pt_2, is the product of an aerodynamic pressure recovery term, Pt_3/Pt_2, and a heat addition pressure recovery term, Pt_4/Pt_3, and these latter two terms are determined from one-dimensional considerations.

The symbols $<$ in the schematic of Fig. 1.4 represent flame holders, which experience a drag force when exposed to a flowing stream and cause a change in stream thrust between stations 2 and 3. In accordance with the requirement for conservation of momentum, the drag force must equal the change in stream thrust:

$$D = P_2A_2(1 + \gamma M_2^2) - P_3A_3(1 + \gamma M_3^2) \tag{1.36}$$

The other elements that cause a drag force can be treated in a similar fashion, and all of the individual element drag contributions can be summed to form a total burner drag that accounts for the total change in stream thrust between stations 2 and 3:

$$D_B = F_2 - F_3 \tag{1.37}$$

A burner drag coefficient can be defined based on the area and dynamic pressure at station 2:

$$C_{D_B} = \frac{D_B}{q_2A_2} = \frac{F_2 - F_3}{q_2A_2} \tag{1.38}$$

Because of wide variations in burner geometry, the burner drag coefficient is not usually a true measure of the actual drag forces on the burner elements, but can be used to represent the total pressure loss model for any desired burner geometry. For a known value of C_{D_B} and known conditions at station 2, the stream thrust per unit area at station 3, $f_3 = F_3/A_3$, can be determined and the corresponding value of M_3 can be found in the tabulated lists of Ref. 2. The actual level of total pressure recovery can then be determined from continuity:

$$\frac{Pt_3}{Pt_2} = (1 + f/a)\frac{\left(\dfrac{P}{Pt}\mathring{m}\right)_2 A_2}{\left(\dfrac{P}{Pt}\mathring{m}\right)_3 A_3}\sqrt{\frac{Tt_3}{Tt_2}} \tag{1.39}$$

where $Tt_3 = Tt_2$ by definition of station 3, conditions at station 2 are known, and M_3 has been determined from stream thrust balance.

In the engine schematic of Fig. 1.4, heat is added in the constant area duct between stations 3 and 4, and in the absence of any drag forces, there is no change in stream thrust:

$$F_3 = F_4 \qquad \text{or} \qquad f_3 = f_4 \tag{1.40}$$

However, the stream Mach number increases with the increase in total temperature, and there is a loss in stream total pressure associated with the increase in entropy. The Mach number at station 4 can be found in the hot deck station of Ref. 2 at the known value of f or f/p. The total pressure recovery can be determined from continuity:

$$\frac{Pt_4}{Pt_3} = \frac{\left(\dfrac{P}{Pt}\,\mathring{m}\right)_3}{\left(\dfrac{P}{Pt}\,\mathring{m}\right)_4}(1 + f/a)\sqrt{\frac{Tt_4}{Tt_3}} \tag{1.41}$$

where $(1 + f/a)\sqrt{(Tt_4/Tt_3)} = (1 + f/a)\sqrt{(Tt_5/Tt_0)}$, which was described in Section 1.4

Generally, the subsonic combustion ramjet engine is analogous to a closed channel with two throats, where the second throat, the nozzle, has a larger area than the first throat, the inlet, to accommodate increases in entropy and temperature between the throats. As the entropy level is changed by variations in fuel flow rate, the flow conditions at any station in the engine will also change, and the distribution of losses across the components will be different. Variations in flow conditions at any station can also be accomplished by changing the nozzle throat size relative to the inlet throat size. Thus, the pressure losses across the combuster are a function of burner geometry as characterized by C_{DB}, fuel flow rate as measured by $(1 + f/a)\sqrt{(Tt_4/Tt_3)}$, nozzle throat size, A_5/A_R, and a Mach number, M_3, at the bookkeeping station, which is dependent on both fuel flow rate and exit nozzle size.

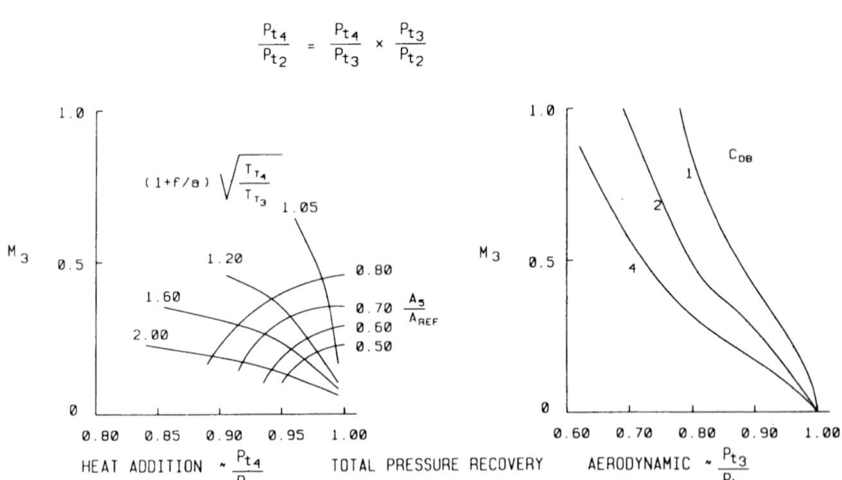

Fig. 1.7 Combustor total pressure recovery as a function of the Mach number at station 3.

Solutions to the set of Eqs. (1.38–1.41) for a range of the independent variables described earlier and selected values of M_3 are plotted on Fig. 1.7. For any value of M_3 the heat addition pressure recovery increases with decreasing nozzle throat size at any fixed value of fuel flow rate and increases with decreasing fuel flow rate at any fixed value of nozzle throat size. For any value of M_3 the aerodynamic pressure recovery increases with decreasing value of burner drag coefficient. An assumption is made in the aerodynamic plot that the value of burner drag coefficient is independent of nozzle throat size. For an increase in nozzle throat size, A_5/A_R, from 0.50 to 0.80, the inlet Mach number M_2 for a typical burner geometry operating at a fixed temperature ratio will increase from 0.11 to 0.16 for a corresponding increase in C_{D_B} from 2.0 to 2.2 and a change in pressure ratio of 0.005. Therefore, the assumption is valid within the limits of engineering accuracy. Experimental data generated during the decade of the 1950s (Ref. 3 is an example) have revealed that most conventional burner configurations operate at a relatively constant value of C_{D_B} over a fairly wide range of heat addition and airflow conditions. A C_{D_B} of 2.0 represents a reasonable level that is attainable with practical burner geometries. The total pressure recovery across a combustor with a C_{D_B} = 2.0 is shown in Fig. 1.8 over a range of heat addition parameter as a function of nozzle throat sizes. The picture presented in Fig. 1.8 shows that combustor total pressure recovery increases with decreasing nozzle throat size, and, for the smaller throat sizes, the pressure recovery is very nearly constant over a wide range of heat addition from near cold flow to fuel rich. Generally, for any fixed nozzle size the aerodynamic losses predominate at the lower levels of heat addition, and the larger the nozzle throat, the higher the losses. As more heat is added, the Mach numbers through the engine decrease; thus, the aerodynamic losses decrease. Even though the heat

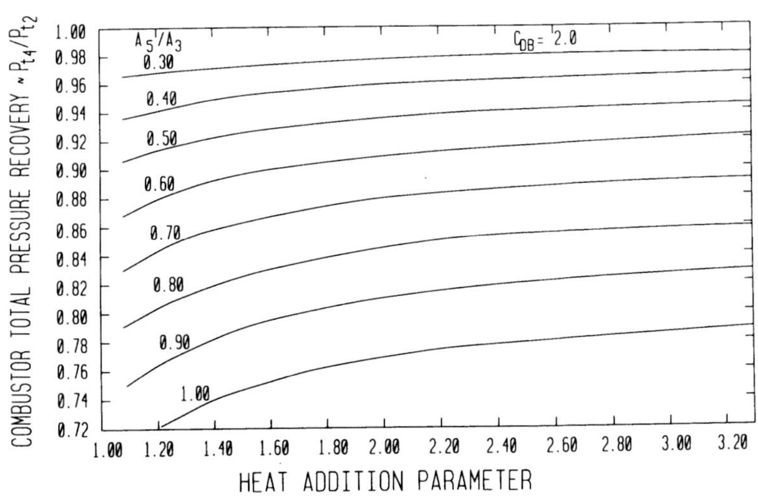

Fig. 1.8 Combustor total pressure recovery for a burner drag coefficient of 2.0.

addition losses increase, the product establishes the trend shown in Fig. 1.8 of increasing total pressure recovery with increasing heat addition for all nozzle throat sizes.

One-dimensional flow techniques are usually inadequate to describe the diffusion process through the inlet component. Combinations of one- and two-dimensional and sometimes three-dimensional techniques are necessary to effect a complete analysis. The higher order of complexity associated with the multidimensional problem warrants a treatise of considerable extent in order to examine the inlet function. Prior to a discussion of such inlet topics as design methods, performance estimating procedures, operational characteristics, performance limiting parameters, installation effects, and test techniques, it is essential to establish boundary conditions on the subject of inlets.

1.6 Inlet Overview

Inlets may be classified in accordance with the following descriptions that are not necessarily all-inclusive and are arranged in an arbitrary order:

1) operating Mach number regime (subsonic, transonic, supersonic, hypersonic, or combinations);
2) family (axisymmetrical, two dimensional, or three dimensional);
3) geometry (fixed or variable);
4) supersonic diffuser form (all external, all internal, or combination external/internal);
5) supersonic compression complexity (normal shock, single surface, isentropic surface, or multiple surfaces);
6) supersonic compression direction [outward (relative to a vehicle body), inward, or downward];
7) location on vehicle body (nose, chin, cheek, top, bottom, side, forward, mid, or aft);
8) number (single, dual, three, or four); and
9) interface with combustor [in-line (usually with podded engines) or off-set (flow dumps from inlet into combustor with integral engines)].

As defined herein, the word *compression* usually refers to the process, whereas the word *diffuser* refers to the structure that constrains the process.

Category 1 may be divided into three familiar classes of operating inlets:

1) Subsonic only with a range from static sea level to about Mach 0.8. Major applications include almost all modern commercial transports (Boeing 7 × 7 series, DC-8, DC-9, and DC-10, Lockheed L1011), the B-52 bomber, the KC-135 tanker, cruise missile, with all of the aforementioned powered by turbojet engines. The inlet consists essentially of a subsonic diffuser, which is relatively simple and well documented with few difficulties and will be discussed only in supersonic applications as a normal shock configuration.

2) Subsonic, transonic, and supersonic with a range from static sea level to about Mach 2.5. Major applications include modern military fighter-

type aircraft (F-14, F-15, F-16, F-18), the Concorde SST, and many bomber aircraft (such as the B-1), with all of the aforementioned powered by turbojet engines. The inlet consists of a supersonic diffuser, throat section, and subsonic diffuser with variable geometry and elaborate controls. Two representative versions of this variable geometry inlet class will be described in Chapter 8.

3) Supersonic only with a range from Mach 2.0 to Mach 4.0. Major applications include older missiles like BOMARC and TALOS, which were operational with the Air Force and Navy, respectively, and proposed new missiles that will be ramjet powered for cruise, but boosted from launch by solid-propellant rocket. The inlet consists of supersonic diffuser, throat section, and subsonic diffuser, is usually a fixed-geometry component, and is less well documented than classes 1 and 2.

The following chapters will be devoted to fixed-geometry inlets for application primarily to unmanned supersonic missiles with diverse physical and operational requirements as determined by both launch platform and target characteristics. Variable-geometry inlets will be discussed as a means for extending the operating range of the ramjet engine powered missile into the hypersonic regime.

2
SUPERSONIC DIFFUSERS

2.1 Compression Process

If the ramjet engine shown in Fig. 1.4 could operate in reverse so that the freestream flow entered the engine at station 6, then the nozzle would compress rather than expand the flow. In the absence of any losses the compression process would be reversible, and the one-dimensional, isentropic area ratio expression [Eq. (1.35)] would apply. The nozzle-shaped inlet could be sized to achieve Mach 1 at the throat plane for any given supersonic freestream Mach number. Unlike the expansion process, however, the compression process is always accompanied by shock waves, which increase the entropy level and cause a loss in total pressure. Furthermore, the action of viscosity produces a different flow picture in the two processes. The decreasing pressures in the flow direction along the expansion surfaces promote attached boundary layers with minimum disruption of the potential flow, whereas the adverse pressure gradient along the compression surfaces can cause boundary-layer separation, which can negate the entire compression process. As a consequence of shock losses in the supersonic diffuser, a throat Mach number of unity is unachievable, and the transition from supersonic to subsonic flow usually occurs across a normal shock wave. Because of viscous action, boundary-layer control is an essential adjunct of inlet design procedures.

The reversed nozzle used in the foregoing discussion represents a conical supersonic diffuser with all internal compression (the process is completely enclosed by solid boundaries) and would require variable geometry to swallow the starting shock and establish supersonic flow throughout the diffuser. This configuration presents a high order of difficulty in both design and development and is near the top on a scale of complexity.

2.2 Compression Complexity

If an inlet designed for subsonic operation were boosted into the supersonic regime, a normal shock wave would form at or upstream from the cowl lip plane and cause the means for transition from supersonic to subsonic flow. The supersonic diffuser in this case would be zero length, all external compression, and fixed geometry with no viscous losses—the ultimate in simplicity. The normal shock wave is a discontinuity in the flow between two regions where the one-dimensional, isentropic relationships shown in Chapter 1 are valid. Changes in flow properties across the normal shock, depicted in Fig. 2.1, can be determined from the requirement for conservation of mass, momentum, and energy per unit area:

$$\rho_1 u_1 = \rho_2 u_2$$

$$P_1 + \rho_1 u_1^2 = P_2 + \rho_2 u_2^2$$

$$\frac{u_1^2}{2} + \frac{\gamma}{\gamma - 1}\frac{P_1}{\rho_1} = \frac{u_2^2}{2} + \frac{\gamma}{\gamma - 1}\frac{P_2}{\rho_2}$$

where ρ is density, u velocity, P pressure, and γ the ratio of specific heats, C_p/C_v.

The subscripts used in the shock-wave equations and the subsequent Prandtl-Meyer equations are consistent with the notation in Ref. 1, where stations 1 and 2 represent conditions upstream from and downstream from the wave, respectively. These equations can be manipulated as shown in Refs. 1 and 4 to form the Rankine-Hugoniot relationships,

$$\frac{P_2}{P_1} = \frac{\dfrac{\gamma + 1}{\gamma - 1}\dfrac{\rho_2}{\rho_1} - 1}{\dfrac{\gamma + 1}{\gamma - 1} - \dfrac{\rho_2}{\rho_1}} \tag{2.1}$$

$$\frac{\rho_2}{\rho_1} = \frac{1 + \dfrac{\gamma + 1}{\gamma - 1}\dfrac{P_2}{P_1}}{\dfrac{\gamma + 1}{\gamma - 1} + \dfrac{P_2}{P_1}} \tag{2.2}$$

and Prandtl's relationship,

$$u_1 u_2 = a^{*2} \tag{2.3}$$

where a^* is the critical speed of sound and is computed from the energy

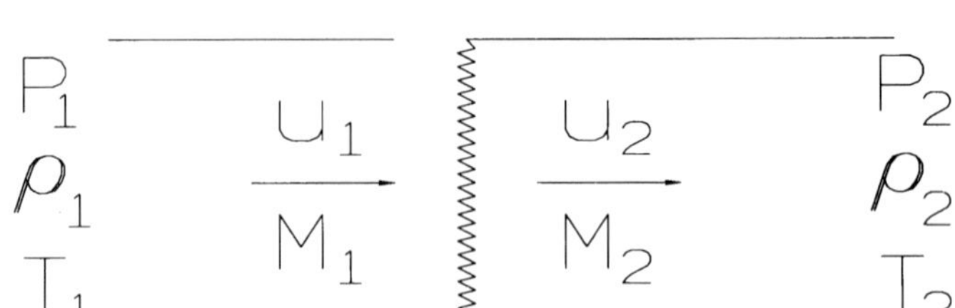

Fig. 2.1 Normal shock-wave nomenclature.

equation written in the form

$$\frac{1}{2}u^2 + \frac{1}{\gamma - 1}a^2 = \frac{\gamma + 1}{2(\gamma - 1)}a^{*2} \tag{2.4}$$

Dividing by u^2, substituting M for u/a, and performing the mathematical steps yields

$$\left(\frac{u}{a^*}\right)^2 = \frac{\gamma + 1}{(\gamma - 1) + (2/M^2)} \tag{2.5}$$

Substituting Eq. (2.5) into Eq. (2.3),

$$M_2^2 = \frac{1 + \dfrac{\gamma - 1}{2}M_1^2}{\gamma M_1^2 - \dfrac{\gamma - 1}{2}} \tag{2.6}$$

which is the relationship of the Mach numbers before and after a normal shock.

The Mach number before the shock, M_1, is always supersonic; i.e., u_1 is greater than a^*, and the Mach number after the shock, M_2, is always subsonic. The converse is not possible because entropy can only increase in accordance with the second law of thermodynamics. Any of the relationships of flow properties across a normal shock wave can be expressed in terms of upstream Mach number, and the following are tabulated in Ref. 1

$$M_2, \quad \frac{P_2}{P_1}, \quad \frac{\rho_2}{\rho_1}, \quad \frac{T_2}{T_1}, \quad \frac{Pt_2}{Pt_1}, \quad \frac{P_1}{Pt_2}$$

as a function of M_1 from 1.0 to 100.0. In particular, the values of total pressure recovery, Pt_2/Pt_1, listed for selected freestream Mach numbers M_1 show a rapidly increasing performance loss as the inlet is accelerated through the transonic into the low supersonic regime, as seen in Table 2.1.

Table 2.1 Variation of normal shock total pressure recovery as a function of freestream Mach number

M_1	Pt_2/Pt_1
1.0	1.00
1.5	0.93
2.0	0.72
2.5	0.50

This performance loss can be mitigated if the freestream velocity can be reduced prior to the onset of the normal shock wave. A reduction in freestream velocity can be achieved by means of an oblique shock, as shown in Fig. 2.2. Obviously, the flow is now two-dimensional with velocity vectors $\boldsymbol{u}$ normal to the shock wave and $\boldsymbol{v}$ parallel to the shock wave. Changes in flow properties across the shock can be determined from the requirements for conservation of mass, momentum, and energy per unit area:

$$\rho_1 \boldsymbol{u}_1 = \rho_2 \boldsymbol{u}_2$$

$$P_1 + \rho_1 \boldsymbol{u}_1^2 = P_2 + \rho_2 \boldsymbol{u}_2^2 \qquad \text{normal}$$

$$\rho_1 \boldsymbol{u}_1 v_1 = \rho_2 \boldsymbol{u}_2 v_2 \qquad \text{tangential}$$

$$\tfrac{1}{2}(\boldsymbol{u}_1^2 + \boldsymbol{v}_1^2) + h_1 = \tfrac{1}{2}(\boldsymbol{u}_2^2 + \boldsymbol{v}_2^2) + h_2$$

Dividing the tangential momentum equation by the mass equation reveals

$$\boldsymbol{v}_1 = \boldsymbol{v}_2$$

or the tangential velocity is unchanged across an oblique shock, and the energy equation becomes

$$\frac{\boldsymbol{u}_1^2}{2} + h_1 = \frac{\boldsymbol{u}_2^2}{2} + h_2$$

Thus, the three conservation equations for oblique shocks are identical to those for normal shocks, and the Rankine-Hugoniot relationships are the same. Because an oblique shock wave acts as a normal shock wave to the perpendicular velocity vector, the normal shock relationships tabulated in Ref. 1 apply to oblique shocks if M_1 and M_2 are replaced by their normal components $M_1 \sin\theta$ and $M_2 \sin(\theta - \delta)$. The oblique shock-wave equations can be developed to determine the relationship between upstream Mach number M_1, flow deflection angle δ, and shock-wave angle θ:

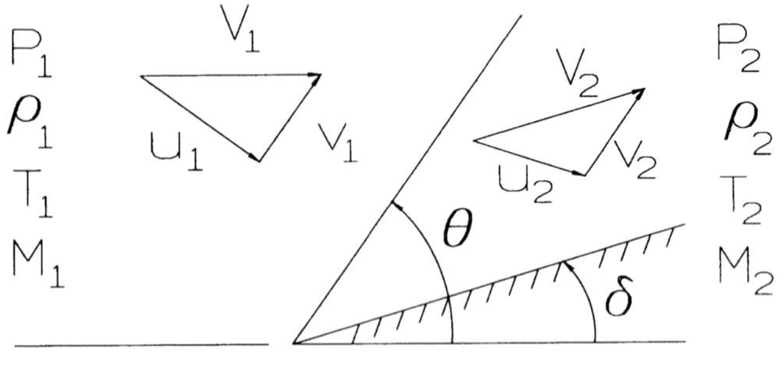

Fig. 2.2 Oblique shock-wave nomenclature.

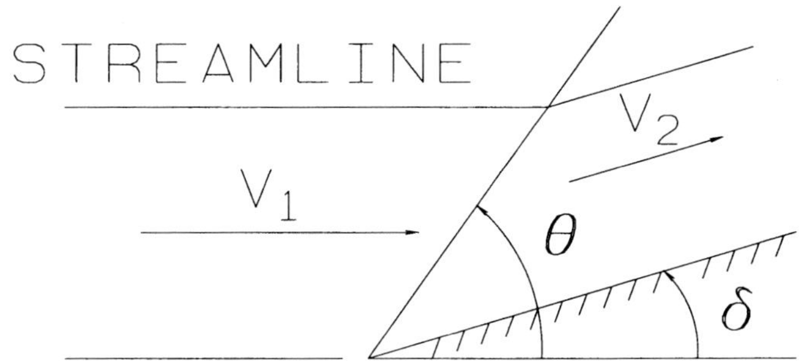

Fig. 2.3 Flow past a wedge.

$$\frac{1}{M_1^2} = \sin^2\theta - \frac{\gamma + 1}{2} \frac{\sin\theta \sin\delta}{\cos(\theta - \delta)} \tag{2.7}$$

For any selected M_1 there is a maximum δ through which the flow can be turned by an oblique shock wave; or, conversely, for a given δ there is a minimum M_1 at which an oblique shock forms. These maximum δ and minimum M_1 form a boundary between shock attachment to or detachment from a solid body that causes the stream deflection. The solid body can be either a wedge or cone, and the flow picture for an attached shock is shown in Fig. 2.3 for the wedge and Fig. 2.4 for the cone. The flow behind the planar wedge shock is uniform and parallel to the wedge surface. The shock wave attached to the cone is also conical in shape, and flow properties are constant on all concentric conical surfaces lying between the shock wave and the body and thus depend only on space variable; i.e., the flow picture is axially symmetrical. The shock attachment boundaries in terms of M_1 and δ for wedges and cones are plotted in Ref. 1.

Solutions to the oblique shock-wave relationships for wedges are also plotted in Ref. 1 where the shock-wave angle θ, the pressure coefficient $(P_2 - P_1)/q_1$, and the downstream Mach number M_2 are shown as a function of the wedge angle δ for lines of constant upstream Mach number M_1. Variations in θ, Pt_2/Pt_1, P_2/P_1, ρ_2/ρ_1, T_2/T_1, M_2, and $\Delta S/R$ are plotted in Ref. 5 as a function of M_1 with lines of constant δ. Although the transition across the conical shock is governed by the oblique shock relationships, the flow undergoes a continuous isentropic compression to surface con-

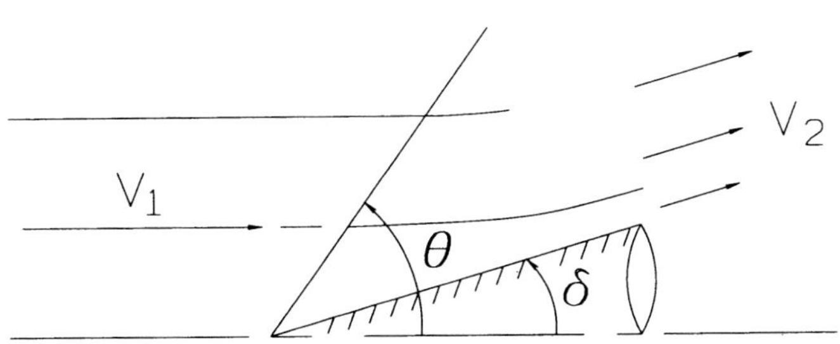

Fig. 2.4 Flow past a cone.

ditions, as shown by the curvature of the streamlines in Fig. 2.4. The flow quantities have been tabulated in Ref. 6, and the solutions are plotted in Refs. 1 and 5.

As an example of the use of the charts presented in the references, the Table 2.2 compares selected flow properties for a single wedge, a single cone, and the normal shock flow process. Inasmuch as the subscript notation of Refs. 1 and 5 is confusing when multiple shocks, such as oblique plus normal, occur in an overall compression process, the station notation introduced in Fig. 1.4 is used henceforth for performance determination with the addition of station 1′ to denote conditions downstream from a normal shock.

At a freestream Mach number of 1.5, the overall recovery of the three processes is on the same order of magnitude, and the simplest would be preferred. At the higher Mach numbers, however, the product of the oblique and normal shocks with the wedge and the cone is considerably higher than the single normal shock recovery, and the added geometrical complexity is beneficial in terms of decreased losses. The cone flow process indicates higher recovery than the wedge, but the average Mach number in the cone flowfield would be higher than the surface value used in the simplified calculations shown in the table, and the average overall recovery for the conical process is usually about the same level as the wedge for comparable strength oblique shocks as measured by the shock angle θ.

Additional performance improvements can be realized by deflecting the flow through a second and third and more oblique shocks. Each additional shock decreases the Mach number at which the normal shock occurs, and the total pressure recovery penalty incurred by each shock is less than that for the preceding deflection. Multiple oblique shock processes can be refined to the ideal limit; i.e., the compression surfaces curves gradually so that

Table 2.2 Total pressure recovery capability of three simple diffuser shapes at three Mach numbers

Flow process	Freestream Mach no. M_0	Oblique shock recovery Pt_1/Pt_0	Surface Mach no. M_1	Normal shock recovery at M_1, $Pt_{1'}/Pt_1$	Overall recovery $Pt_{1'}/Pt_0$
Single	1.5		Detached shock		0.93
wedge	2.0	0.95	1.44	0.95	0.90
$\delta = 15°$	2.5	0.93	1.87	0.78	0.73
Single	1.5	0.99	1.00	1.00	0.99
cone	2.0	0.97	1.42	0.95	0.92
$\delta = 25°$	2.5	0.93	1.76	0.83	0.77
Normal	1.5	1.00	1.5	0.93	0.93
shock	2.0	1.00	2.0	0.72	0.72
	2.5	1.00	2.5	0.50	0.50

the deflection occurs across a number of Mach lines rather than across abrupt discontinuities. The gradual compression process is isentropic (outside the surface boundary-layer regions) and can be incorporated with either a wedge or a conical leading edge. Flow over an initial wedge plus an isentropic surface is shown in Fig. 2.5. The Mach lines that form the compression fan coalesce into a single focal point. Each of the Mach lines in the fan is a small disturbance where the flow vector normal to the line has a magnitude equal to the local speed of sound a. The flow picture used to describe the process is usually one where the Mach lines are expansions, because the equations can be rigorously developed. The resulting equations can be used to approximate the compression process, if the entropy increase is small.

Figure 2.6 shows the flow relationships across a weak expansion wave. The upstream velocity $\boldsymbol{u}_1$ is split into the normal vector $\boldsymbol{u}_N$, which is equal to the local speed of sound a and a tangential vector $\boldsymbol{u}_t$, which is unchanged in magnitude in crossing the wave:

$$\boldsymbol{u}_N = a = \boldsymbol{u}_1 \sin\theta \tag{2.8a}$$

or

$$\sin\theta = a/\boldsymbol{u}_1 = 1/M_1 \tag{2.8b}$$

The oblique angle for a weak wave is a function only of the upstream Mach number, and the angle is called the Mach angle, which is given the special notation μ in Ref. 1, wherein values of μ are tabulated for the range of Mach numbers from 1 to 100.

The downstream flow conditions are represented by three velocity diagrams that highlight different aspects of the flow picture. Figure 2.6a illustrates that there is no change in the tangential velocity vector, whereas the normal velocity vector is increased by an amount d$\boldsymbol{u}_N$. Figure 2.6b shows the increase in the magnitude of the absolute velocity vector from $\boldsymbol{u}_1$ to $\boldsymbol{u}_2$ and the change in direction by an amount dν. Figure 2.6c portrays a differential velocity diagram that is useful in developing mathematical relationships. From Fig. 2.6c,

$$\mathrm{d}\boldsymbol{u} = \mathrm{d}\boldsymbol{u}_N \sin\theta \tag{2.9}$$

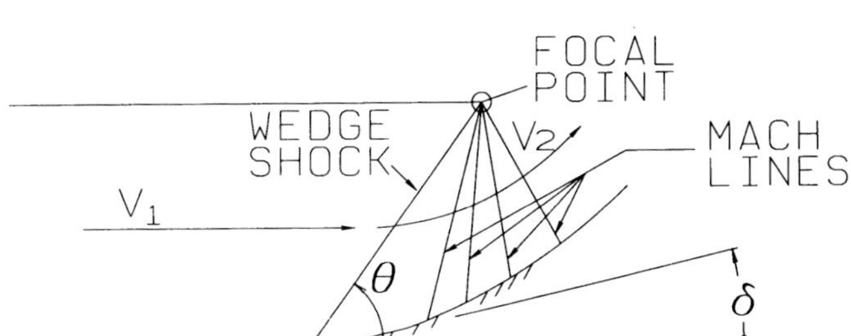

Fig. 2.5 Flow past a wedge and isentropic surface.

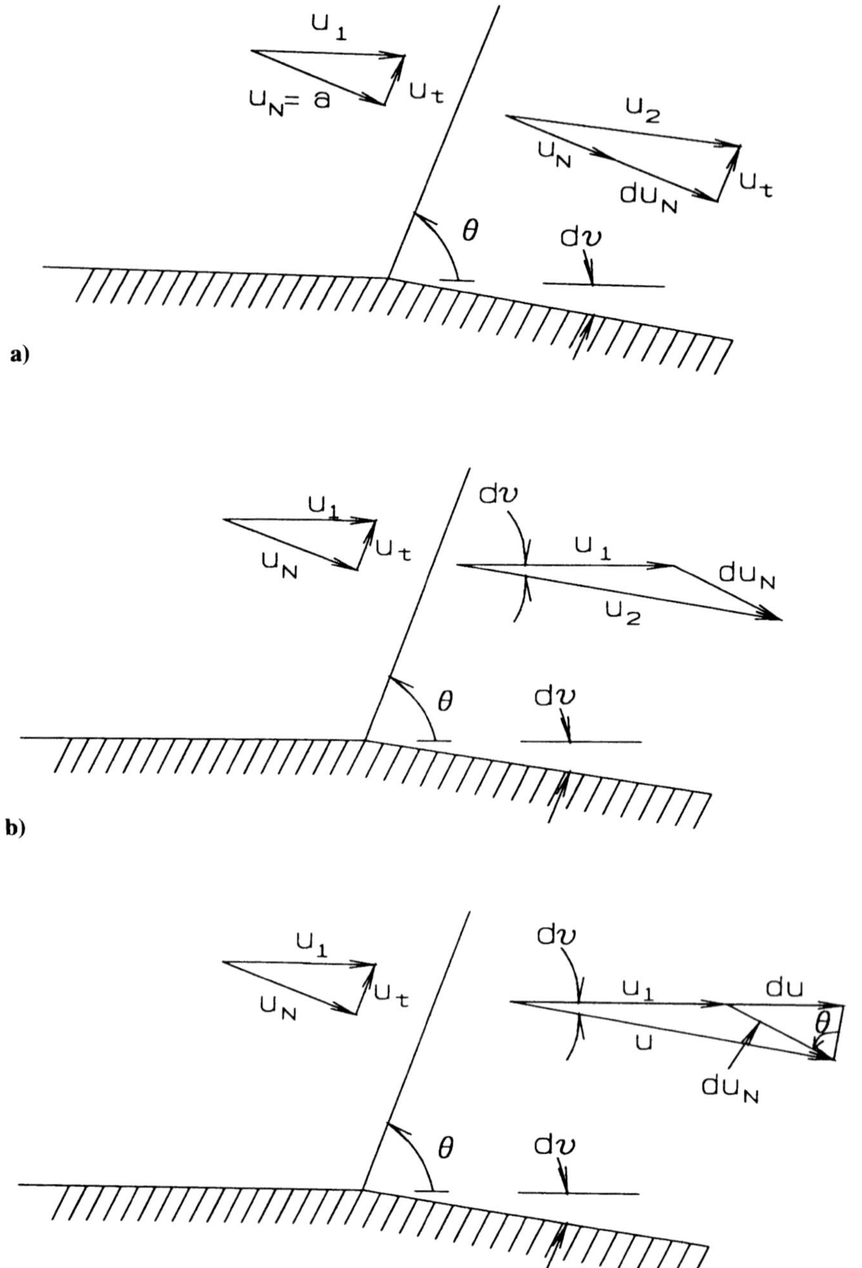

Fig. 2.6 Flow across a weak expansion wave with downstream flow conditions represented by three velocity diagrams: a) unchanged tangential velocity vector; b) changes in magnitude and direction of initial velocity vector; c) incremental changes in velocity vectors.

and

$$\sin d\nu = d\nu = \frac{d\boldsymbol{u}_N \cos\theta}{\boldsymbol{u}} \tag{2.10}$$

Substituting Eq. (2.9) into Eq. (2.10),

$$d\nu = \frac{\cos\theta}{\sin\theta}\frac{d\boldsymbol{u}}{\boldsymbol{u}}$$

and using Eq. (2.8) and the identity $\sin^2\theta + \cos^2\theta = 1$,

$$d\nu = \sqrt{M^2 - 1}\,\frac{d\boldsymbol{u}}{\boldsymbol{u}} \tag{2.11}$$

The variables M and $\boldsymbol{u}$ are related through the energy equation

$$\left(\frac{\boldsymbol{u}}{a^*}\right)^2 = \frac{\gamma + 1}{(\gamma - 1) + (2/M^2)} \tag{2.5}$$

or

$$M^2 = \frac{2(\boldsymbol{u}/a^*)^2}{(\gamma + 1) - (\gamma - 1)(\boldsymbol{u}/a^*)^2} \tag{2.12}$$

Substituting Eq. (2.12) into Eq. (2.11) and showing the integration limits,

$$\int_{\nu_1}^{\nu_2} d\nu = \int_{(\boldsymbol{u}/a^*)_1}^{(\boldsymbol{u}/a^*)_2} \left[\frac{\left(\dfrac{\boldsymbol{u}}{a^*}\right)^2 - 1}{1 - \dfrac{\gamma - 1}{\gamma + 1}\left(\dfrac{\boldsymbol{u}}{a^*}\right)^2}\right]^{\frac{1}{2}} \frac{d\left(\dfrac{\boldsymbol{u}}{a^*}\right)}{\dfrac{\boldsymbol{u}}{a^*}}$$

If $\nu_1 = 0$ at $(\boldsymbol{u}/a^*)_1 = M_1 = 1.0$, the relationship between flow angle and velocity is

$$\begin{aligned}\nu = {} & \frac{1}{2}\sqrt{\frac{\gamma + 1}{\gamma - 1}}\left\{\frac{\pi}{2} - \sin^{-1}\left[\gamma - (\gamma - 1)\left(\frac{\boldsymbol{u}}{a^*}\right)^2\right]\right\} \\ & - \frac{1}{2}\left\{\frac{\pi}{2} + \sin^{-1}\left[-(\gamma + 1)\left(\frac{a^*}{\boldsymbol{u}}\right)^2\right]\right\}\end{aligned} \tag{2.13}$$

Equation (2.13) may be transformed to obtain the expression used in Ref. 1:

$$\nu = \sqrt{\frac{\gamma + 1}{\gamma - 1}} \tan^{-1} \sqrt{\frac{\gamma - 1}{\gamma + 1} (M^2 - 1)} - \tan^{-1} \sqrt{M^2 - 1} \tag{2.14}$$

Equations (2.13) and (2.14) describe the expansion process for a two-dimensional supersonic stream flowing over a convex bend approximated by a series of small angle changes, $\Delta\nu$, in a curved boundary surface, as shown in Fig. 2.7. The expansion is called Prandtl-Meyer flow, and the Prandtl-Meyer angle ν is tabulated in Ref. 1 over a range of Mach numbers from 1.0, where ν is 0 deg to 100, where ν is 127.59 deg. (The maximum possible angle at $M = \infty$ for $\gamma = 1.4$ is 130 deg+.) If the flow quantities are known at any point in the flowfield shown in Fig. 2.7, the properties at any other point can be determined from the change in flow angle between the two points. The same relationships can also be used in the determination of flow properties for the compression of a two-dimensional stream over a concave surface (Fig. 2.5) if the curvature is gradual enough to preclude any significant entropy increase.

Equation (2.11) is also an exact solution of the equations of motion after transformation from the physical (x,y) plane into the velocity $(\boldsymbol{u},\boldsymbol{v})$ plane, which is called the hodograph. The hodograph transformation is described in detail in Refs. 4 and 7. If the integral of Eq. (2.11) as given in Eq. (2.13) is plotted as a polar diagram with velocity $(\boldsymbol{u}/a^*)$ measured radially from the center and angles of flow direction $\Delta\nu$ measured as angles about the center, the result is the plot shown in Fig. 2.8, which is the hodograph of flow over a convex surface. The curve that defines the magnitude of the velocity vectors is called the characteristic for the flow. Construction of the flow along the wall can be accomplished graphically in a step-by-

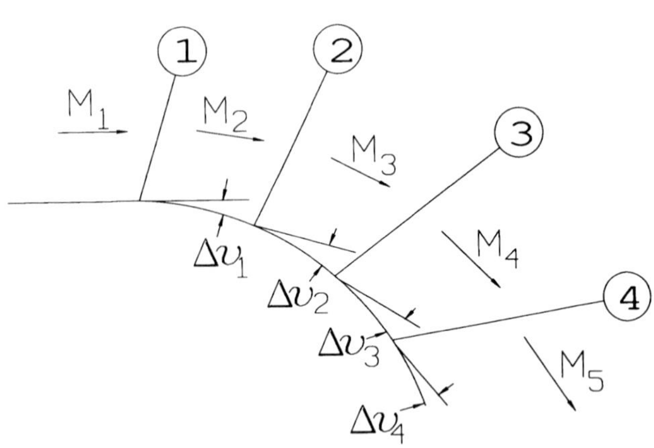

Fig. 2.7 Flow over a convex surface.

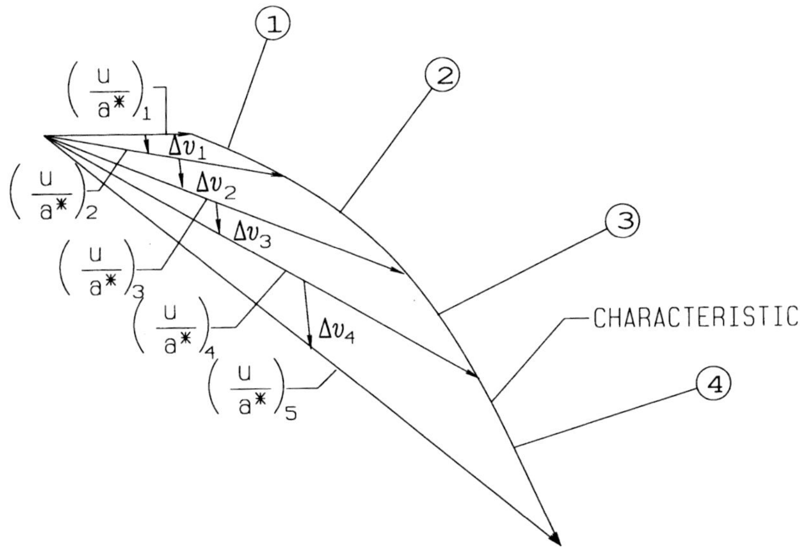

Fig. 2.8 Hodograph of flow over a convex surface.

step process that alternates between the physical plane and the hodograph plane.

1) The wall is divided into a number of finite segments inclined at angles $\Delta\nu_1$, $\Delta\nu_2$, $\Delta\nu_N$ to the horizontal.
2) The hodograph is oriented so that $(\boldsymbol{u}/a^*)_1$ is parallel to the freestream velocity vector $\boldsymbol{M}_1$.
3) The second velocity vector in the hodograph, $(\boldsymbol{u}/a^*)_2$, is drawn from the origin parallel to the first inclined surface (i.e., at angle $\Delta\nu_1$).
4) The Mach line separating the two velocities is determined by drawing the normal to the characteristic, identified as 1, and constructing the Mach line 1 parallel to this line in the physical plane.
5) The process is continued through the rest of the wall segments.

The hodograph diagram of Fig. 2.8 can be enlarged to include characteristic curves for a wall curving upward or for flow between two walls with Mach wave intersections and reflections from solid or free boundaries. Families of characteristic curves provide a network, as shown in Fig. 2.9, which is called a characteristics diagram. An example of the use of this diagram as a hodograph for the flow between two diverging walls (Fig. 2.10) is discussed later. The walls are divided into a number of straight line segments that generate Mach lines. These lines intersect each other and reflect off the opposite wall and in the process form small quadrilateral flowfields wherein flow properties are constant. The region of uniform, parallel upstream flow, labeled 1 in Fig. 2.10, transforms into a single point, also labeled 1, in the hodograph plane. The flow expands into regions 2 and 2′ in accordance with the picture of the flow over a single convex surface described earlier and maps into points 2 and 2′ on the hodograph. The equations of motion are satisfied across the Mach lines that separate regions 2 and 2′ from 1. Region 3 lies downstream from the intersection of the two Mach lines, and the equations of motion can be satisfied only if point 3 in the hodograph lies on the intersection of the characteristic curves that contain points 2 and 2′. To preserve symmetry, the flow in region 3 must have the same direction but higher magnitude than that in region 1. The process is continued until the entire field is mapped in both planes. The accuracy of the solution may be increased by dividing the walls into smaller increments. This flow-mapping process may also be used to solve cases in which finite changes of entropy occur, i.e., a compression field as illustrated by Fig. 2.11. If a strong shock occurs in the field, as with a 15-deg-angle leading-edge wedge, the solution can be computed with a combination of the shock polar (oblique shock properties as presented in Ref. 1) and the characteristics diagram. Intersections of strong shock waves and weak Mach waves or curved shocks can produce entropy changes in the flow regions downstream from the intersection, and these regions are separated by slipstreams or vortex sheets, parallel to the streamlines. The flow on each side of the vortex sheet must have the same direction

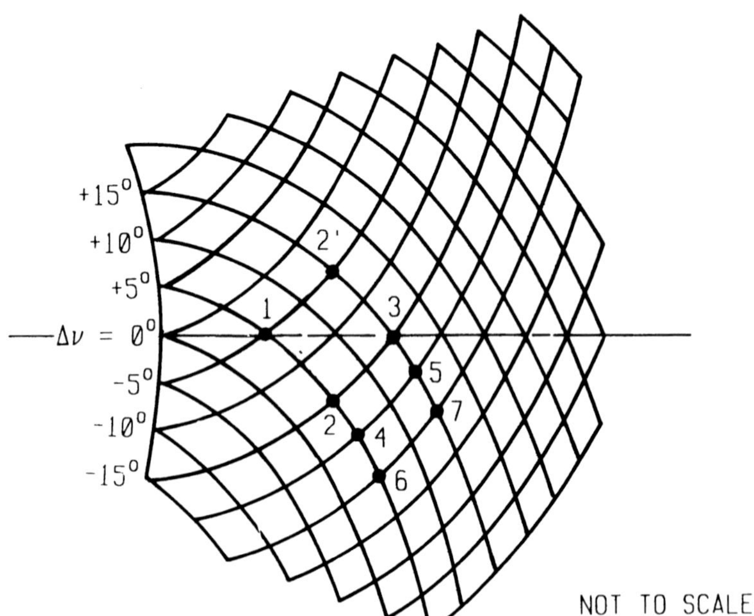

Fig. 2.9 Characteristics diagram where numbered points illustrate hodograph of flow between two diverging walls.

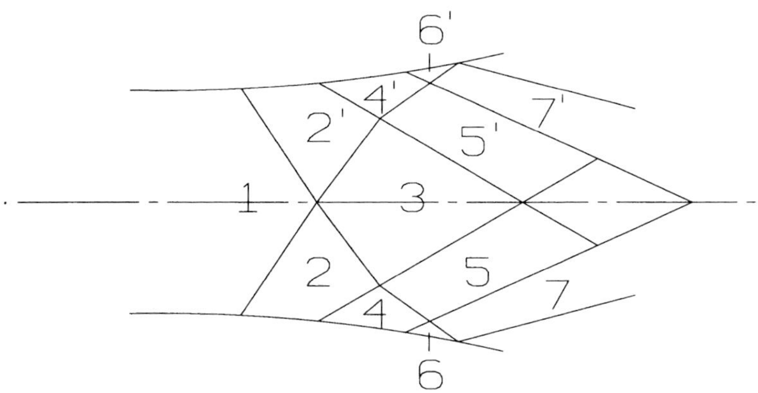

Fig. 2.10 Flow between two diverging walls.

and the same static pressure, even though all other stream properties may have different values.

The characteristics solution for an open nose body of revolution with an internal cone (i.e., the reversed exit nozzle used as the opening example in this chapter) is presented in Ref. 7. One significant aspect of the flow picture is that the leading-edge shock increases in intensity with decreasing radius and becomes normal at the axis, and this phenomenon is general for supersonic flow in circular channels. Thus, there is a total pressure defect at the centerline of this type of diffuser.

2.3 Diffuser Shapes

Regardless of the complexity of the compression process, a diffuser of practical (buildable) shape can usually be designed to accomplish the compression with acceptable efficiency. A representative cross section of diffuser shapes keyed to the inlet classifications delineated in Chapter 1 is shown in the sketches in Fig. 2.12. The stations used to determine supersonic diffuser performance are given on the side view sketches, and these stations are consistent with the notation in Fig. 1.4 and the 1′ station added in Table 2.2. The simplest configuration (i.e., the normal shock)

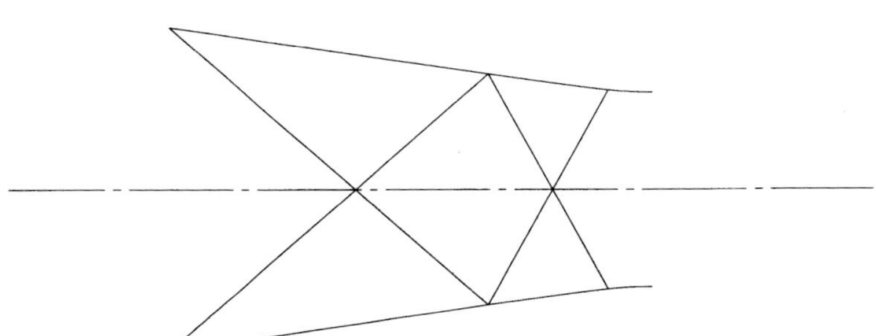

Fig. 2.11 Flow between two converging walls.

can have almost any imaginable cross-sectional shape, including round, square, or elliptical (all three shown on the figure) or half-round, rectangular, triangular, etc. The shock occurs in, or slightly upstream from, the cowl lip plane of the inlet, which is the minimum area, or throat, and is normal to the freestream flow direction. The single-surface configuration diverts the freestream flow to a new direction and reduces the magnitude of the velocity vector by means of an oblique shock. The normal shock still occurs in or near the cowl lip (throat) plane, but is skewed in a plane normal to the flow direction downstream from the oblique shock. The maximum flow deviation consistent with oblique shock attachment to the single surface at a freestream Mach number of 2.0 is 40.6 deg for a half-angle cone and 23 deg for a wedge. The flow is compressed from a freestream shape of half a circle into an annular shape at the cowl lip plane, from a square into an equal width but much smaller height rectangle, or from a sugar scoop into an ellipse. The isentropic surface usually consists of a relatively low-angle leading-edge cone or wedge followed by a sufficient number of degrees of gradual (shock-free) turning to achieve more compression of the stream tube than was possible with the single cone or wedge, or scoop. Theoretically, the flow could be turned through the Prandtl-Meyer limit ($\nu \sim 130$ deg+), but there is a practical isentropic turning limit for acceptable performance. The limiting turning angle is plotted in Fig. 2.13 as a function of freestream Mach number M_0. The sketch in the figure shows the flow picture that exists at the turning limit. Essentially, two streams of air with a different compression history must share a common boundary formed by the vortex sheet downstream from the focal point. Inasmuch as the static pressure and flow direction on each side of the sheet must be identical, the two streams must adjust either by accelerating or diffusing. The limit is reached when the upper stream must be subjected to a normal shock at freestream Mach number in order to balance conditions. Details of the construction of the curve are given in Ref. 8.

Multiple-surface configurations can be formed by two opposing wedges (or sugar scoops) of equal conformation that aid each other in compressing the flow through intersecting and reflecting weak shock waves into a rectangular (or elliptical) throat. In both configurations the side walls are flat and cut back, or notched, so that flow can be spilled immediately upstream from the throat station, if the inlet should unstart. The top four configurations in Fig. 2.12 compress the flow in a manner that is called all external; i.e., the flow channel upstream from the minimum area plane or section is open on one or more sides. If a disturbance should cause the normal shock to move upstream from the throat (inlet unstart), the open channel permits immediate flow spillage to stabilize the shock in a position where the inlet total pressure recovery is practically unchanged from the critical value. When the cause of the disturbance is removed, the inlet can swallow the shock (restart) automatically, and there is no need to vary any aspect of the geometry.

However, variable geometry is required for the two configurations shown in the bottom of Fig. 2.12. In one case some of the compression is accom-

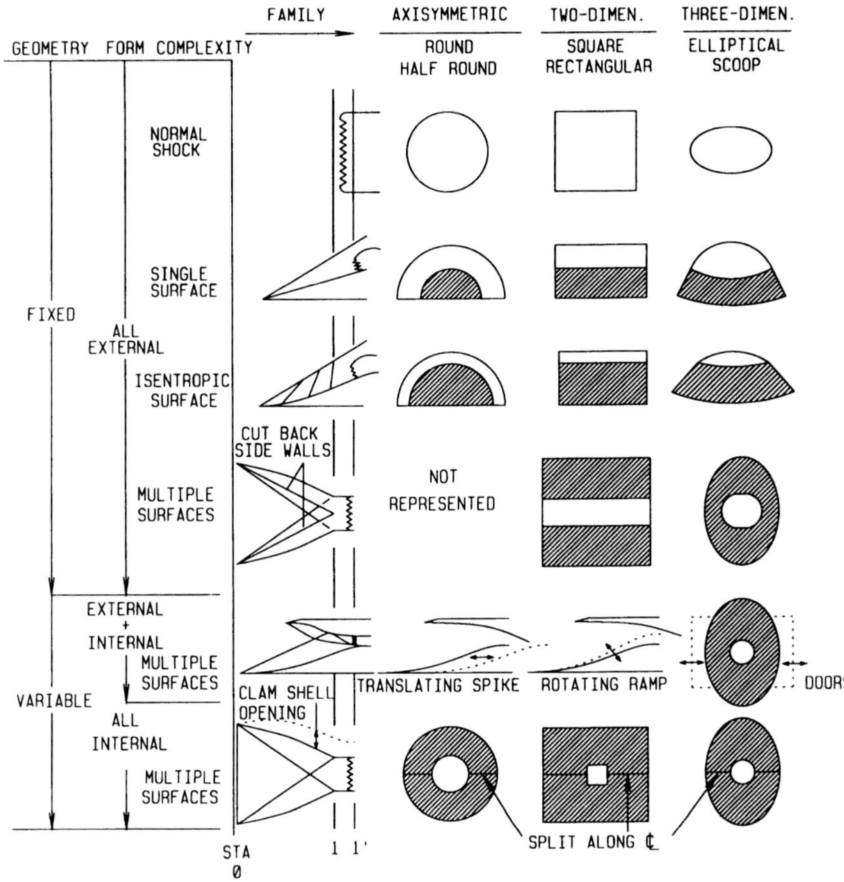

Fig. 2.12 Representative supersonic diffuser shapes for a variety of inlet classifications.

plished externally, but a significant amount of the compression is accomplished internally; i.e., the process is completely enclosed by solid walls. In the other case all of the compression is internal between the entrance station and the throat station. In both cases multiple surfaces work in conjunction to compress the flow through a series of compression waves with intersections and reflections into a throat that has a much smaller area than the entrance area. The ratio of the entrance area A_0 to the throat area A^* is called the contraction ratio of the inlet. If the inlet unstarts because of a disturbance generated externally (e.g., wind gust) or internally (combustor oscillation), the normal shock must move from the throat station to a station just upstream from the entrance to the enclosed channel in order to spill air. In order to effect a restart (swallow the shock), the throat area must be large enough to pass the entrance area mass flow at

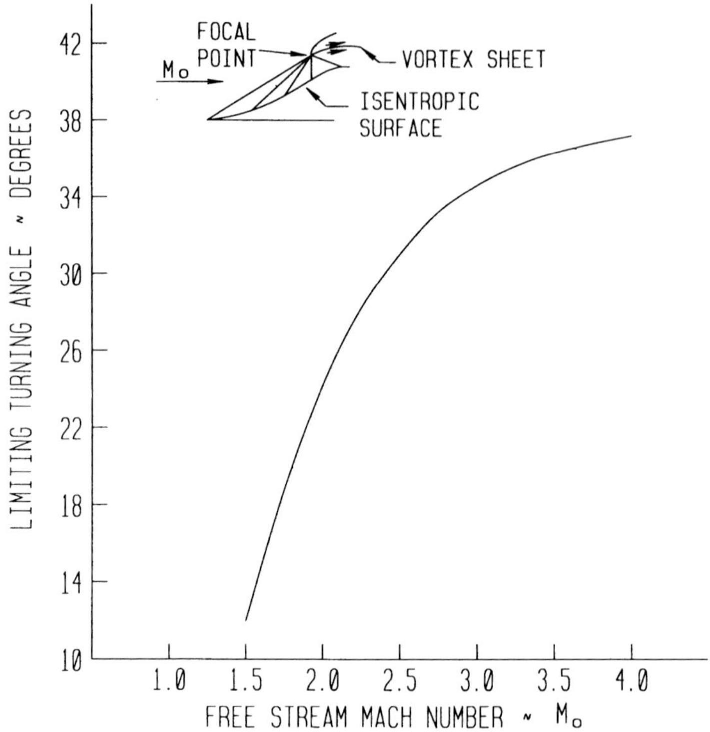

Fig. 2.13 Maximum isentropic compressive turning.

a total pressure that corresponds to the value behind a normal shock at the entrance Mach number:

Physically,

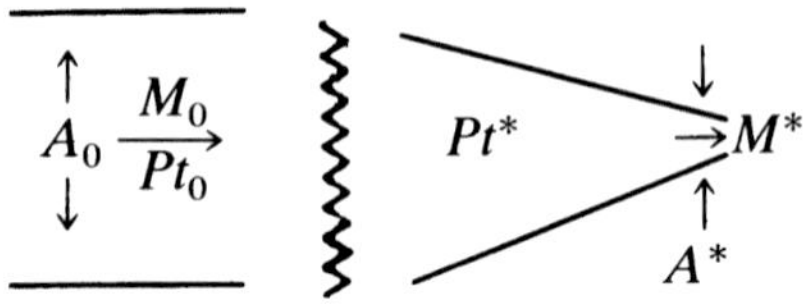

Mathematically, the equation for conservation of mass requires

$$\frac{Pt_0 \left(\frac{P}{Pt} \mathring{m}\right)_0 A_0}{\sqrt{Tt_0}} = \frac{Pt^* \left(\frac{P}{Pt} \mathring{m}\right)^* A^*}{\sqrt{Tt^*}} \tag{2.15}$$

where $Tt_0 = Tt^*$ and Pt^*/Pt_0 is the total pressure ratio across a normal shock at M_0. Also, $[(P/Pt)\mathring{m}]$ is a function only of Mach number, as shown in Ref. 2.

Rearranging:

$$\frac{A_0}{A^*} = \frac{Pt^*}{Pt_0} \frac{\left(\frac{P}{Pt}\mathring{m}\right)^*}{\left(\frac{P}{Pt}\mathring{m}\right)_0} \tag{2.16}$$

The term A_0/A^* is called the starting contraction ratio and is plotted in Fig. 2.14 vs freestream Mach number M_0. If the contraction ratio of the supersonic diffuser exceeds the value plotted at a specified Mach number, the inlet will be able to start only with the aid of some variable-geometry device. If the inlet could be started and an isentropic compression process established ($Pt^*/Pt_0 = 1.0$), Eq. (2.16) becomes

$$\frac{A_0}{A^*} = \frac{\left(\frac{P}{Pt}\mathring{m}\right)^*}{\left(\frac{P}{Pt}\mathring{m}\right)_0} \tag{2.17}$$

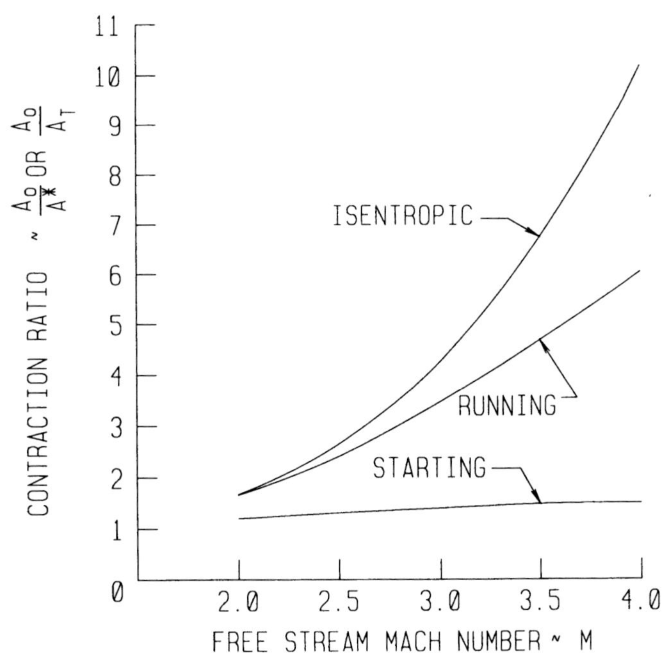

Fig. 2.14 Significant contraction ratios for supersonic diffusers.

which is the same as Eq. (1.35) and is the expression for the isentropic contraction ratio wherein the flow can be compressed from any freestream Mach number to Mach 1 in the throat. The isentropic contraction ratio is also plotted in Fig. 2.14 and represents a theoretical limit for the reversible adiabatic compression of an inviscid gas with a γ of 1.4. A practical contraction ratio limit has been empirically determined on the basis of a multitude of inlet tests conducted during the last 30 years. The curve labeled "running" contraction in Fig. 2.14 represents an average of measurements gathered by numerous industrial concerns and government agencies in many different wind tunnels with models of varying sizes and shapes. Some of the variable-geometry schemes devised to transform a supersonic diffuser from the starting geometry to the running geometry are illustrated by the bottom two sets of configurations in Fig. 2.12. For the external/internal case half of an axisymmetric inlet is shown with a spike that translates axially from a forward position for starting to a retracted position for running. A two-dimensional inlet is depicted with a hinged ramp that is rotated from a full open-throat position for starting to a much smaller throat opening for running. Also, a three-dimensional inlet that compresses the flow from an elliptical capture shape into a circular throat has doors that are open during inlet start and closed for running. The all-internal configurations feature a split centerline with half the diffuser hinged to open like a clam shell for starting. In the running position the compression process for all three configurations is three dimensional, and the shocks generated by the compression surface leading edges will be curved because of interactions with shocks in orthogonal planes. The flow picture can be solved by three-dimensional characteristics for rotational flow. Reference 7 describes the numerical calculations used to determine the internal flow properties of the conical diffuser shown as the example configuration under the axisymmetric family heading in Fig. 2.12.

The total pressure recovery across the supersonic diffuser ($Pt_{1'}/Pt_0$) affords a measure of the performance capability achievable with different levels of compression complexity represented by the sketches in Fig. 2.12. Five different total pressure recovery curves are plotted against freestream Mach number in Fig. 2.15. Each curve is a locus of peak performance points for a series of supersonic diffusers of comparable complexity designed at innumerable Mach numbers within the range of 2.0–4.0. The normal shock configuration, the least complex, has the lowest recovery at all Mach numbers. The single-surface configurations, as represented by the 15-deg wedge (or the 25-deg half-angle cone, or the sugar scoop), with one level of complexity above the normal shock, achieves a higher recovery than the normal shock. The isentropic surface configuration, with a 5-deg initial wedge angle (or a 15-deg half-angle initial cone) followed by gradual turning to a final angle that is several degrees below the turning limit shown in Fig. 2.13, has the pressure recovery level identified by the middle curve of the five shown in Fig. 2.15. This "conservatively constrained" configuration yields a balanced compromise between internal performance, as measured by total pressure recovery, and external performance, as measured by drag, with a shape that is not unduly complex. If the isentropic

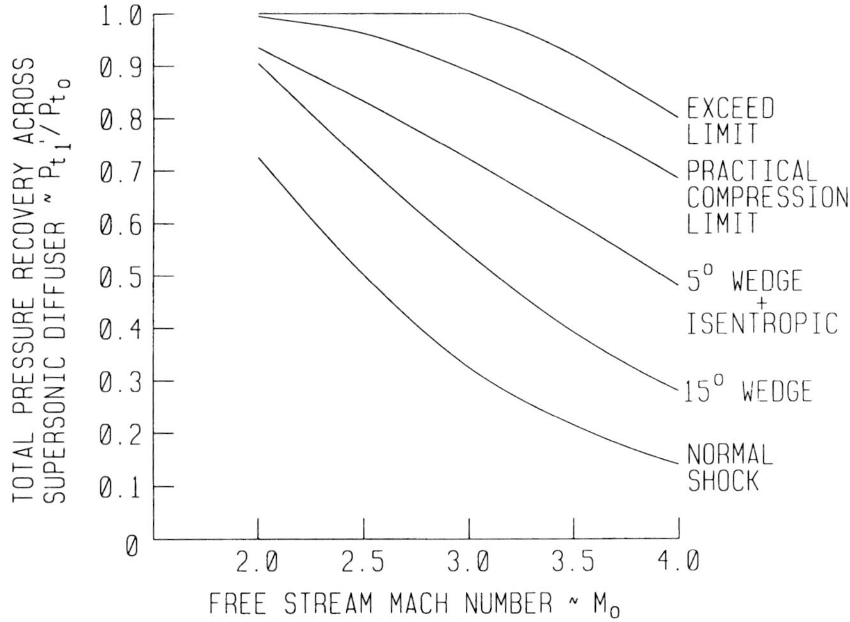

Fig. 2.15 Total pressure recovery capability of supersonic diffusers designed at five different levels of compression complexity.

turning is continued to the limiting angles defined by Fig. 2.13, the fourth curve, designated "practical compression limit" results, but the identical total pressure recovery curve is also produced by multiple-surface configurations with significant internal compression capability operating at the running contraction ratio shown in Fig. 2.14. Thus, the same supersonic compression limit would seem to apply regardless of whether the process is accomplished externally or internally. The compression limit is not an impenetrable barrier, and many diffusers have been designed to exceed the turning limit in the case of external compression configurations and the contraction limit in the case of internal compression configurations. Generally, all of the limit-exceeding diffusers are impractical, particularly above the design point, because of excessive drag penalties, severe reductions in the mass flow available for combustion, and/or unacceptable restrictions on the useful operating range. The fifth and topmost curve in Fig. 2.15 shows the locus of total pressure recovery values measured with supersonic diffusers designed to exceed the external compressive turning limit[9,10] and supersonic diffusers that have exceeded the internal running contraction limit by means of the application of massive amounts of boundary-layer bleed up to 25% of the total mass flow entering the diffuser.[11,12]

In the Mach number range of 2.0–4.0, which is the range of primary interest herein, the normal shock inlet is not competitive on the basis of total pressure recovery capability and will not be discussed further except as a point of comparison. For application to compact missiles, all variable-geometry diffusers, although capable of attaining the highest possible recovery levels, are too complex (also costly, voluminous, and heavy) and will not be described further, except that the external/internal configurations will be used as a means for extending missile operating capability into the hypersonic regime.

Most of the text henceforth will be devoted to fixed-geometry components of the axisymmetric and two-dimensional families, because these inlet classifications have the most extensive data base (both ground and flight) and offer the widest choice of relatively proven configurations for application to current, conventional vehicles. For some applications, where such factors as drag reduction, radar cross section, and heat-transfer requirements are paramount, the three-dimensional family would contain the preferred configuration. In a particular application with severe limitation on maximum heat release, the low external drag of the multiple-surface double-scoop supersonic diffuser with cut-back side walls was the overriding factor in the choice of this three-dimensional configuration over a variable-geometry axisymmetric inlet with higher total pressure recovery. The double-scoop inlet also has the advantage of flexibility in the shape of the cross sections at the capture and throat stations. The capture shape can be an ellipse with any ratio of major to minor axis. Similarly, the throat can be elliptical or circular, and the throat shape can be translated in the vertical direction any amount with the restriction that the throat shape must be contained within the outline of the capture shape. Furthermore, the capture ellipse can be wrapped to conform to the vehicle body surface, so that the projected frontal area of the total configuration can be minimized. Some of the techniques used in the design of the double scoop diffuser are presented at the end of the next section, and the performance achieved with a series of inlets that feature the double scoop diffuser is given in Chapter 6.

2.4 Design Procedures

All supersonic diffusers are designed so that the aerodynamic contours produce peak performance at a specific freestream Mach number. For applications where the missile is boosted to cruise Mach number and there are no requirements for acceleration, the supersonic diffuser (or inlet) design Mach number is readily apparent. However, as the interval between end of boost and cruise Mach numbers increases, the choice of design Mach number becomes more difficult. Trajectory analyses have shown that a design Mach number near the midpoint of the interval usually offers the best compromise between accelerating and cruise performance requirements. After the design Mach number has been selected, the inlet is sized to capture the exact amount of freestream air required in the combustion process for a specified engine thrust level. Installation constraints dictate the number of inlets used to capture the mass flow. For each inlet of the axisymmetric family, the cowl radius is determined by the area of the stream tube that must be captured. For two-dimensional inlets where the height-to-width ratio can be varied on either side of unity, most flight-test data have been acquired for square configurations. Reductions in inlet length, weight, and parasitic side-wall surface area are achieved by selecting inlets with a height-to-width ratio less than unity. Thus, the critical capture dimension is the maximum inlet width allowed by configurational constraints.

For single-surface configurations the pressure recovery level of the supersonic diffuser is controlled by the wedge angle in the case of the two-dimensional family and by the cone half-angle in the case of the axisymmetric family. At any freestream Mach number, as the wedge (cone) angle increases, the oblique shock recovery decreases, but the wedge (cone surface) Mach number decreases, so that the normal shock recovery increases. The product of the oblique shock recovery and normal shock recovery at the wedge (cone surface) Mach number is shown in Fig. 2.16. All of the curves peak at differing values of wedge (cone) angle, and all of the curves have an upper limit defined by shock detachment from the wedge (cone). The best internal performance is achieved at the wedge (cone) angle commensurate with the peak value of the curve at the selected design Mach number. However, experience has shown that a balanced compromise between internal performance as measured by total pressure recovery and external performance as determined by cowl drag level is attained over a wide range of freestream Mach numbers with a wedge angle of 15 deg and a conical half-angle of 25 deg.

As soon as the design Mach number, cowl height or radius, and wedge or cone angle have been selected, the on-design flow picture for the single-surface supersonic diffuser with all external compression can be readily constructed, as shown in Fig. 2.17. The cowl lip is located on the oblique shock so that the capture area of the inlet A_C equals the freestream area A_0 to be delivered to the combustor. For the two-dimensional configuration the Mach number everywhere in the triangular field bounded by the wedge surface, the oblique shock, and the normal shock has a constant magnitude and a flow direction parallel to the wedge surface. The inner cowl surface is aligned with this flow direction, and the leading edge of the outer cowl surface has the lowest possible included angle (usually about 5 deg) consistent with structural integrity. Both cowl surfaces are turned back toward the horizontal in as short a distance as practical to minimize the external surface exposed to pressure drag forces. Most two-dimensional supersonic diffusers have parallel side walls to constrain the flow within the triangular field. For the axisymmetric configuration the Mach number in the field bounded by the cone surface, the oblique shock, and the normal shock has a magnitude that varies from a minimum at the cone surface to a maximum behind the oblique shock wave and a flow direction with maximum deflection from the freestream value at the cone surface to minimum deflection behind the oblique shock. The inner cowl surface is aligned with the latter flow direction, and the other features of the cowl are determined as described earlier for the two-dimensional configuration. For configurations with a flow-aligned cowl, both the internal and external cowl surfaces must have increasingly higher angles relative to the freestream flow direction as the wedge (cone) angle increases. Thus, increasing cowl drag

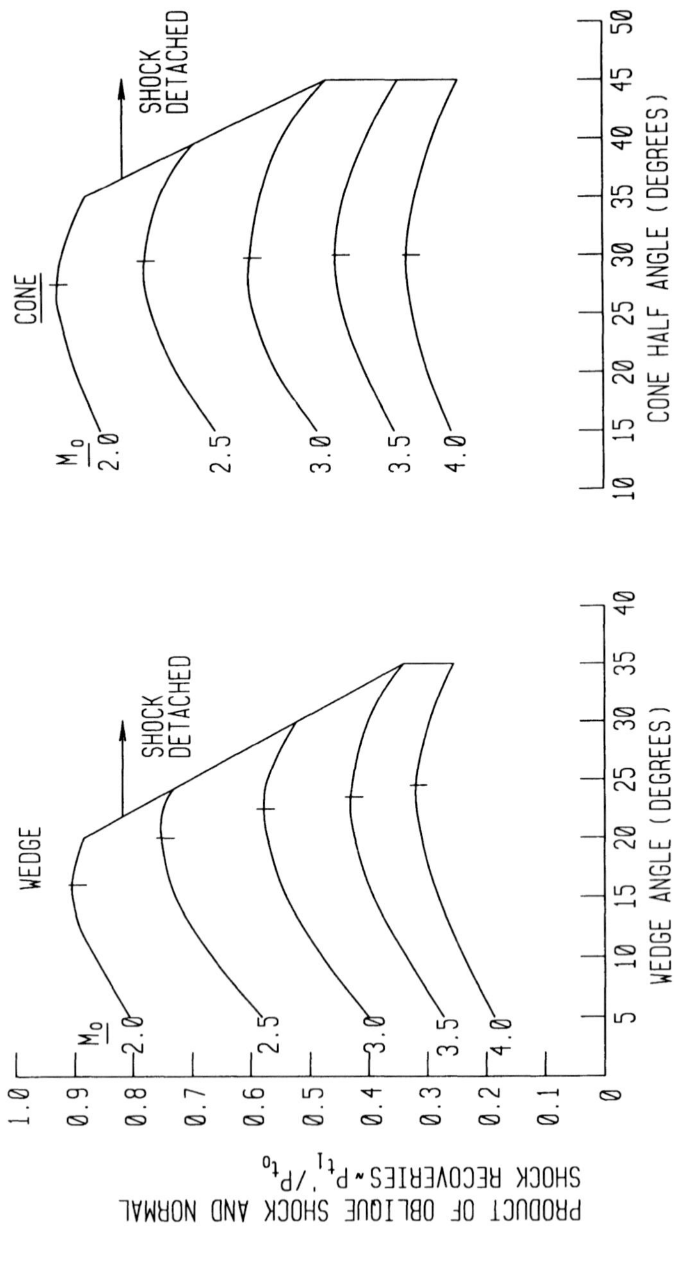

Fig. 2.16 Total pressure recovery capabilities of supersonic diffusers with single-surface wedges and cones for a multitude of surface angles.

is inevitably a consequence of increasing total pressure recovery, unless the cowl is misaligned, as illustrated in Fig. 2.18. The cowl is flattened relative to the freestream flow direction so that the external surface area exposed to pressure drag is decreased, but the inner cowl surface is oriented at an angle of attack to the wedge (cone) flow. The supersonic wedge (cone) flow is turned to a direction parallel with the inner cowl surface by an oblique shock wave that extends across the passage downstream from the cowl lip plane. It is appropriate to utilize the extra compression afforded by a misaligned cowl by incorporating internal contraction into the supersonic diffuser design. For this purpose the throat plane must be located downstream from the cowl lip plane, and the throat area is determined from the requirement for conservation of mass between the cowl lip plane and throat plane. However, to ensure the self-starting capability necessary for a fixed-geometry component the throat area is limited by the starting contraction ratio curve shown in Fig. 2.14. Typical relationships between internal contraction area ratios and flow-misalignment angles are given in Table 2.3 for a representative case of a 20-deg wedge diffuser operating at a design Mach number of 2.5. The nomenclature used in the table is consistent with that shown in Fig. 2.18.

The starting contraction area ratio of 1.13 for an upstream Mach number of 1.64 is taken from the curve in Fig. 2.14 and represents a limit on the internal contraction permissible for self-starting. Cowl misalignments of 2 deg and 5 deg from the 20-deg wedge flowfield yield contraction area ratios, based on continuity, less than the limiting value, and indicate a "safe"

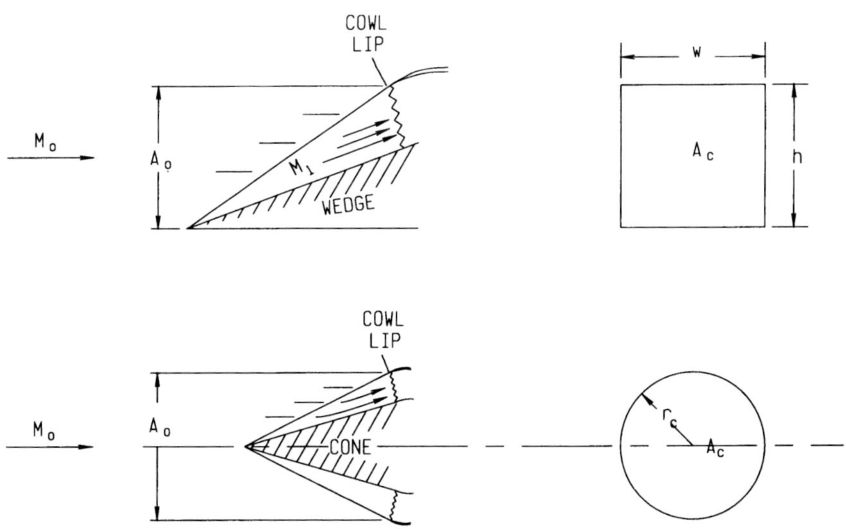

Fig. 2.17 On-design flow picture for two single-surface supersonic diffusers with flow-aligned cowls and all-external compression: a) two dimensional; b) axisymmetric.

Table 2.3 Relationship between cowl misalignment and internal contraction area ratio for a 20-deg wedge diffuser at a freestream Mach number of 2.5

Free-stream Mach number M_0	Wedge Mach number M_1	Starting contraction area ratio (A_1/A^*)	Cowl mis-align angle	Inner cowl surface angle δ_{COWL}	Cowl surface Mach number M_c	Continuity equation parameters $\frac{Pt_c}{Pt_1}$	×	$\left(\frac{P}{Pt}\mathring{m}\right)_c$	÷	$\left(\frac{P}{Pt}\mathring{m}\right)_1$	=	Contraction area ratio A_1/A_t
2.5	1.64	1.13	2°	18°	1.58	1.000		.43080		.41430		1.04
			5°	15°	1.46	.998		.46235				1.11
θ Wedge			7°	13°	1.39	.995		.47929				1.15
= 20°			10°	10°	1.28	.990		.50257		↓		1.20

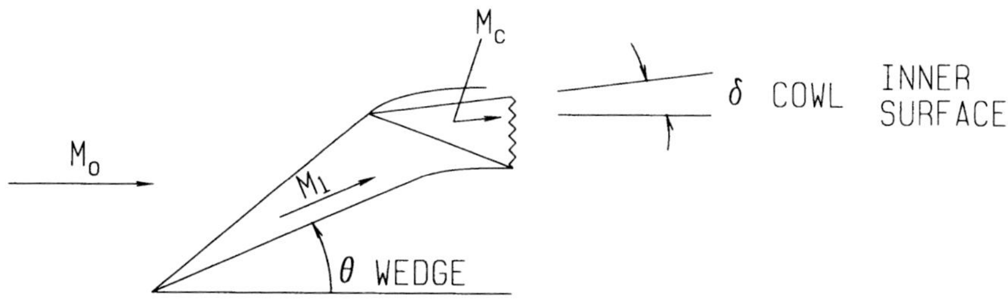

Fig. 2.18 On-design flow picture for single wedge surface supersonic diffuser with misaligned cowl and some internal contraction.

design. Cowl misalignments of 7 and 10 deg exceed the starting limit. The configuration with a 5-deg cowl misalignment has about a 7% increase in total pressure recovery compared with the aligned cowl configuration, and the external cowl drag is reduced by virtue of both a lower surface angle and a lower projected frontal area.

For fixed-geometry components the strictures imposed by the self-starting requirement severely limit the contribution of internal compression process so that the walls of the enclosed passage do not have an opportunity to interact with one another. Figure 2.18 shows the internal cowl lip shock impinging on the wedge surface at the throat station. There are no wave reflections or intersections of waves from the two walls, and the rather simple design approach illustrated by Table 2.3 is adequate to describe the internal contraction portion of a fixed-geometry supersonic diffuser.

The basic design procedures described earlier for the single wedge (cone) surface supersonic diffuser with or without internal contraction can be applied to the isentropic surface configuration, except that the pressure recovery level is controlled by the amount of compression surface turning from the freestream flow direction. It is possible to design a pure isentropic turning surface with an infinitely thin leading edge and no shock loss, but such a configuration is impractical. Generally, a semi-isentropic surface with a 5-deg leading-edge wedge or a 15-deg leading-edge half-angle cone combines low total pressure loss with adequate structural integrity. Definition of the isentropic surface coordinates and the flow properties at any point in the compression field generated by the curved surface can best be determined from the method of characteristics, which can be programmed on mainframe computers to yield the physical plane solution shown in Fig. 2.19. Although a rectangular two-dimensional diffuser with an initial 5-deg wedge is portrayed in the figure, the flow picture is representative of any single semi-isentropic surface, where all of the Mach lines that comprise the compression fan are focused at a single point on the leading-edge shock. Generally, the cowl lip is located on the focal point at the design Mach number, and the inner surface of the cowl is aligned with the flow direction at the completion of the isentropic turning.

The characteristics solution begins with the determination of the wedge shock angle θ and the magnitude of the Mach number M_w in the wedge flowfield. The initial Mach line is constructed across the flowfield from the

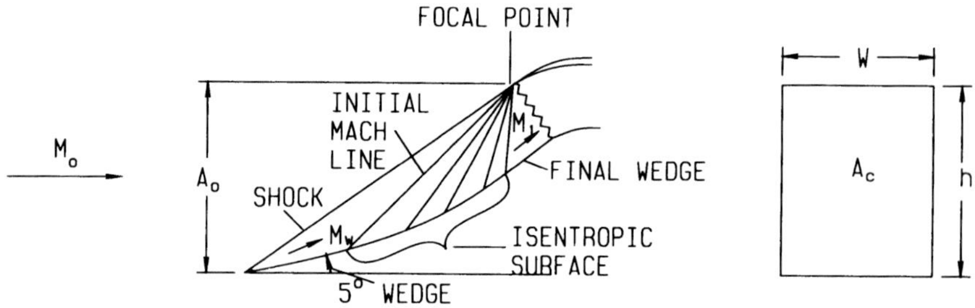

Fig. 2.19 On-design flow picture for semi-isentropic surface supersonic diffuser with flow-aligned cowl.

end of the straight wedge surface to the intersection with the shock wave. For the example chosen herein (a program developed at The Marquardt Co., Van Nuys, CA), the initial line is called the first left line (LL 1) and is divided into 40 points, which represent intersections with right running characteristic lines (RL). The program printout lists the x and y coordinates of each point, the Mach number, the flow angle, the static temperature, and the ratio of static to total pressure. The coordinates are nondimensionalized in that the length of the wedge in the axial dimension, x, is one unit and the height of the wedge, y, is the tangent of the wedge angle. Some of the details of the isentropic flowfield construction are shown in Fig. 2.20 to illustrate the network. For clarity more than half of the lines and half of the points on each line have been omitted. The example program constructs a left line at each half of a degree of turning and construction always begins at the focal point. Each left line ends at the compression surface point, which is determined by interpolation between left line intersections with the two proximate right lines. For fine enough nets, mass and momentum are automatically conserved. With each succeeding left line the flow area becomes smaller, and there are fewer points between the focus and the surface. The last left line in Fig. 2.20 has a constant value of the x coordinate and is commensurate with a final flow angle of 35 deg at a Mach number of 1.221. (For an axisymmetric net Mach number and other flow properties would vary between the focal point and the surface.) As shown in Fig. 2.13, the isentropic turning limit at a freestream Mach number of 2.5 is 31 deg, which is achieved at left line (LL) no. 53 in the net, and the corresponding Mach number is 1.368. A conservatively constrained diffuser, in accordance with the definition of the middle curve in Fig. 2.15, would be one designed for a final surface angle of 20 deg (LL 31) and a final Mach number of 1.743. In each of the three design cases (conservative turning, compression limit turning, above limit turning), the cowl lip is anchored on the focal point, and the inner cowl surface is aligned with the final flow direction. For flattened cowls the misalignment angle must be added to the compression surface turning angle to measure total turning. The cowl surface replaces the vortex sheet shown in Fig. 2.13. Because the solid cowl can sustain a pressure differential between inner and outer surface and differing flow directions, the normal shock shown

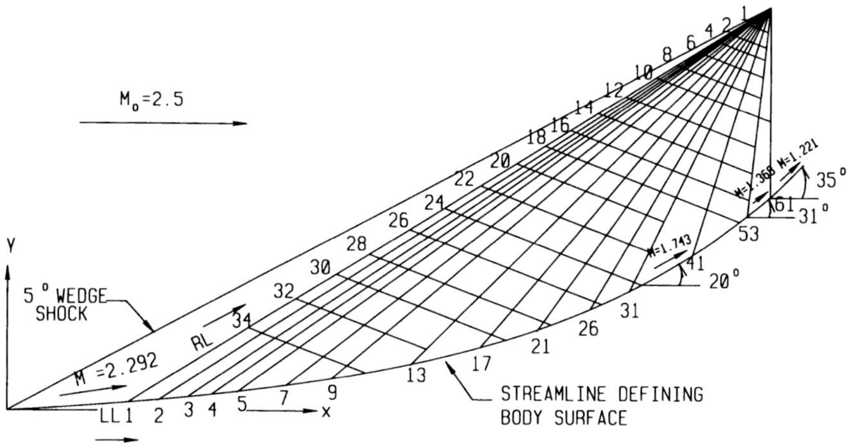

Fig. 2.20 Example of a characteristics net used in the design of a semi-isentropic surface.

in Fig. 2.13 will not form, and the compressive turning limit can be exceeded without the performance penalty associated with the flow picture in Fig. 2.13. However, if the focal point should move inside the cowl during some off-design operating condition, the flow picture of Fig. 2.13 and the attendant performance degradation will apply.

Aerodynamic contours for many members of the three-dimensional family of supersonic diffusers can be determined by tracing streamlines through either two-dimensional or axisymmetric characteristics nets. Although these streamline tracing techniques may be less accurate than three-dimensional numerical computational solutions, the principles are easier to comprehend and implement. In the particular case of a double-scoop supersonic diffuser (Fig. 2.21 is an example of a typical configuration), the streamline tracing method produces contours commensurate with expected performance results as determined experimentally in scaled model wind-tunnel tests. The method is applicable for any conformation where the throat shape is contained within the capture shape, and the diffuser is symmetrical about the vertical (Y) axis. The capture shape can be expressed mathematically:

$$Y = \frac{b}{a}\sqrt{a^2 - Z^2} + CZ^2 \tag{2.18}$$

and the throat shape

$$Y = G \pm \frac{R_b}{R_a}\sqrt{R_a^2 - Z^2} \tag{2.19}$$

where

Y = ordinate for plane normal to X axis
Z = abscissa (see Fig. 2.21)

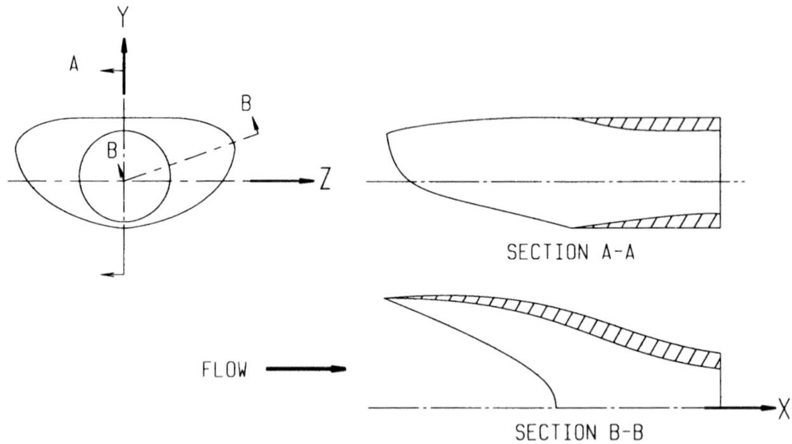

Fig. 2.21 Sketch of a typical double-scoop supersonic diffuser.

a = major axis of capture ellipse
b = minor axis of capture ellipse
C = factor to determine amount of wrap of ellipse
R_a = major axis of throat ellipse or circle if $R_a = R_b$
R_b = minor axis of throat ellipse or circle if $R_a = R_b$
G = factor to determine amount of throat translation

The relationship between the area of the capture shape and the area of the throat shape is determined by the specified value of the contraction ratio:

$$\mathrm{CR} = \frac{A_C}{A_t} \tag{2.20}$$

For a self-starting diffuser with no variable devices, the contraction ratio cannot exceed the value given in Fig. 2.14 for the selected design Mach number. If the desired total pressure recovery level requires a contraction ratio higher than the starting value, some form of variable geometry, such as the doors shown in Fig. 2.12, will be essential.

The diffuser is sliced by a number of flow planes (about 20 if symmetrical about the Z axis, otherwise, 40) in an arrangement as shown in Fig. 2.22, where the ratio of capture area to throat area between any two adjacent planes must equal the contraction ratio. Thus,

$$\frac{A_{1234}}{A_{1265}} = \mathrm{CR}$$

It is expedient at this point to define a polar coordinate system using the symbols shown in Fig. 2.23 for an example flow plane, where B is the

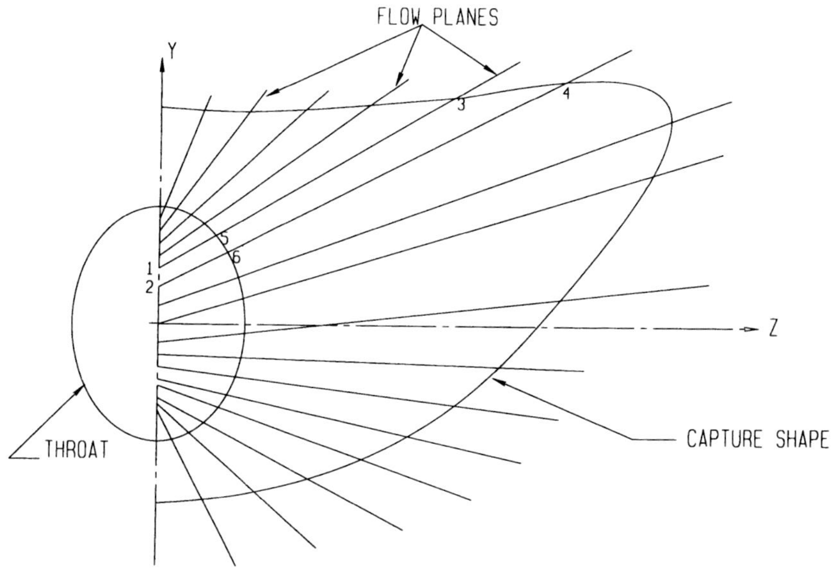

Fig. 2.22 A flow plane arrangement through half of a double-scoop supersonic diffuser.

ordinate of the flow plane intercept on the Y axis; w is the flow plane inclination relative to the Z axis; and y is the radial or slant distance from the intercept.

The flow plane axis a-b (perpendicular to the Y-Z plane in Fig. 2.23) is a line on the surface of a cylinder so that the flow in any plane is annular. The Mach number distribution or static pressure gradient along all axes a-b from the freestream to the throat are selected to be identical, so that there will be no crossflow in the vertical or Y plane. The upper part of the flow on the vertical plane ($Z = 0$, $w = 90$ deg) is axisymmetric about a point in the vertical plane so that the radius of the cylinder (K) for this point is zero. The value of B for this flow plane can be determined from

$$\left(\frac{Y_s + B}{Y_t - B}\right)^2 = \text{CR} \tag{2.21}$$

and the radius of the cylinder (K) for each flow plane can be determined from

$$\frac{(y_s + K)^2}{(y_t + K)^2} - \frac{K^2}{K^2} = \text{CR} \tag{2.22}$$

In each flow plane an axisymmetric characteristics net can be generated within the boundary conditions established by a cylindrical surface of known radius, a uniform and parallel freestream flow at a given Mach number,

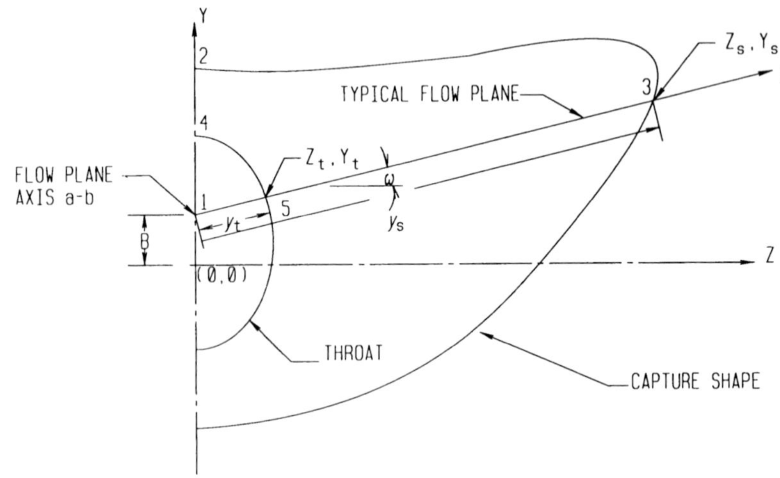

Fig. 2.23 Example flow plane showing polar coordinate symbols.

and an undistorted throat flow at a Mach number determined from the contraction ratio. Additional inputs are known values of radial distance in the flow plane from intercept at the $Z = 0$ axis to the capture shape intercept (y_s) and the throat shape intercept (y_t); and the Mach number distribution along the cylindrical surface ($a - b$) in the X plane. Figure 2.24 illustrates the trace of a streamline from the entrance to the throat. It is recommended that at least 20 points be used on the freestream and throat characteristic lines to achieve acceptable accuracy. The streamline in each plane represents an aerodynamic contour with an isentropic leading-edge, which is impractical. A leading-edge wedge with an included angle on the order of 5 deg can replace the leading portion of the isentropic surface with the advantages of shortening the diffuser while providing structural

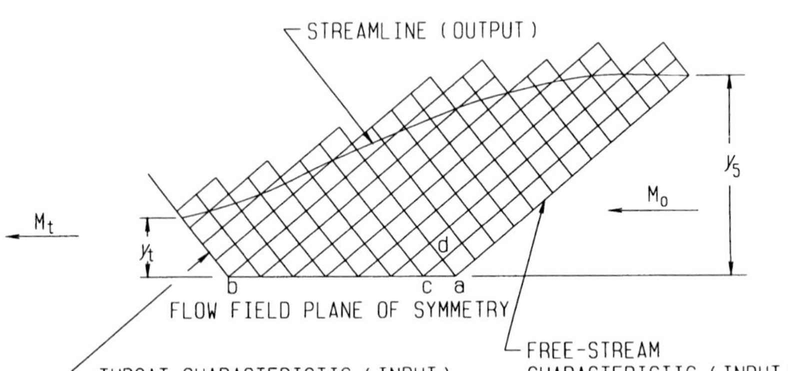

Fig. 2.24 Characteristics net in a typical flow plane.

rigidity and with only a relatively small penalty in total pressure recovery. The wedge angle can vary from plane to plane.

The diffuser surfaces defined in the flow plane coordinates can be transformed into Cartesian coordinates as follows: $X = x$, $Y = B + y\,(\cos w)$, and $Z = y(\sin w)$. More complete details of the streamline tracing technique are contained in Ref. 13.

3
SUBSONIC DIFFUSERS

3.1 Compression Process

During operation in the subsonic velocity regime, an air-breathing engine requires an inlet that consists only of a subsonic diffuser. The diffusion of a subsonic flow is accomplished in a diverging channel, or passage, as shown in Fig. 3.1 in the schematic of a conical diffuser that is representative of the process in a diffuser of any desired shape with the same length L, flow areas at entrance and exit, and, therefore, the same, or equivalent, conical half-angle θ. If the area change per unit length, $d(\pi r^2)/dx$, of the conical diffuser is low enough to ensure a thin boundary layer with a high kinetic energy profile along the walls, the compression process will be very nearly isentropic and can be accurately analyzed by means of the one-dimensional techniques described in Chapter 1. Experiments conducted with subsonic diffusers of all shapes with an overall area change, $d(\pi r_2)/dL$, limited by an equivalent conical half-angle of 3 deg, have shown a compression process total pressure recovery, Pt_2/Pt_1, approaching unity. Compression of a uniform, parallel, freestream flow at a Mach number of 0.8 to a diffuser exit Mach number as low as 0.2 has been accomplished in a conical diffuser with total pressure recovery on the order of 0.97 and with a relatively undistorted exit flow.

However, during operation in the supersonic velocity regime, where the overall compression process must be accomplished in two stages, the supersonic diffuser most often delivers a distorted main flow profile and an asymmetric boundary layer that interacts adversely with the terminal normal shock. This departure from the "one-dimensional" subsonic diffuser-alone picture complicates the analysis of the compression process and causes additional losses. Whereas the supersonic diffuser can be designed with inviscid procedures as developed in Chapter 2, the subsonic diffuser is designed with empirical procedures that must necessarily incorporate analytically intractable shock/boundary-layer interactions.

3.2 Throat Section

The transition from supersonic to subsonic flow in any inlet usually occurs across a normal shock that stands in the minimum area, or throat, under

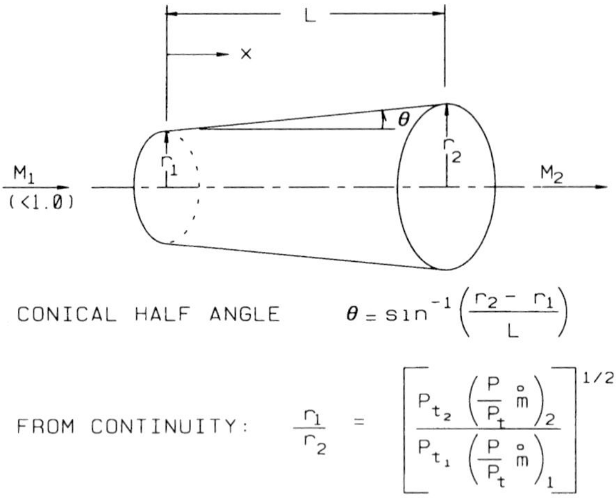

Fig. 3.1 Schematic of a conical subsonic diffuser.

normal design operating conditions. In an inviscid flow process the normal shock can stand as a single wave, or discontinuity, as discussed in Chapter 2. Because of viscosity, which will be discussed in Chapter 4, boundary layers grow along the wall, or walls, of the supersonic diffuser, and the normal shock interacts with these boundary layers to form a series of discontinuities, or a shock train, as shown in the sketch in Fig. 3.2. The presence of the shock train in the subsonic diffuser adds to the existing pressure gradient, dP/dx, so that the boundary layer separates from the subsonic diffuser wall or walls. The flow picture that accompanies boundary-layer separation is also sketched in Fig. 3.2, and the results are a decrease in the amount of subsonic diffusion, a substantial reduction in the process total pressure recovery, and a severely distorted flow profile that can degrade combustion efficiency.

A solution to the problem is described in Ref. 14, in which the throat of the inlet is stretched to contain the entire shock train. The sketch in Fig. 3.3 shows the shock/boundary-layer interaction confined to the throat section with attached boundary layers along the subsonic diffuser walls and a relatively undistorted flow profile. The authors of Ref. 14 also investi-

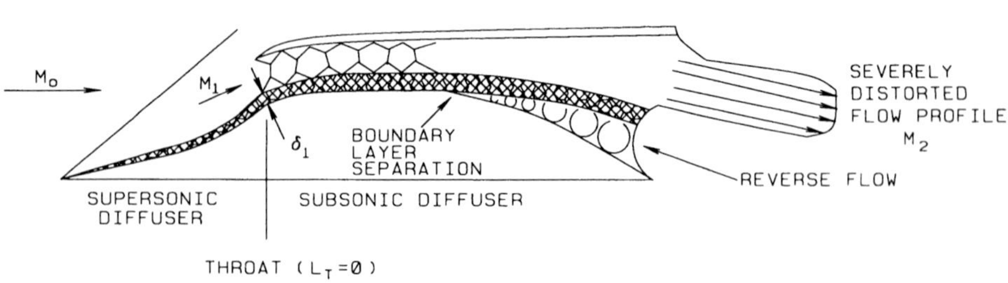

Fig. 3.2 Normal shock interaction with a boundary layer in a subsonic diffuser.

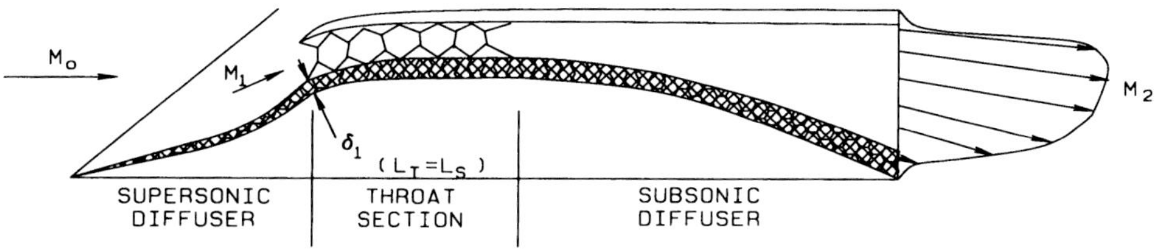

Fig. 3.3 Normal shock interaction with a boundary layer in a constant-area throat section.

gated the effect of changes in throat length relative to shock train length, and these results are summarized in Fig. 3.4. If the throat length is equal to the shock train length, the total pressure recovery of the compression process is the maximum attainable with the inlet configuration and is close to the pressure recovery across a single normal shock at the inlet entrance Mach number. If the physical throat length is longer than the shock train length, there are additional viscous losses, but the overall total pressure recovery decreases only gradually with increasing throat length. If throat length is shorter than shock train length, the total pressure recovery at the highest Mach number tested in Ref. 14 decreases very rapidly with decreasing throat length, and the flow picture at zero throat length must be similar to that shown in Fig. 3.2. At the lowest Mach number tested in the reference, the total pressure recovery decreases less rapidly with decreasing throat length, and at zero throat length the loss in total pressure recovery is considerably less than that in the highest Mach number case.

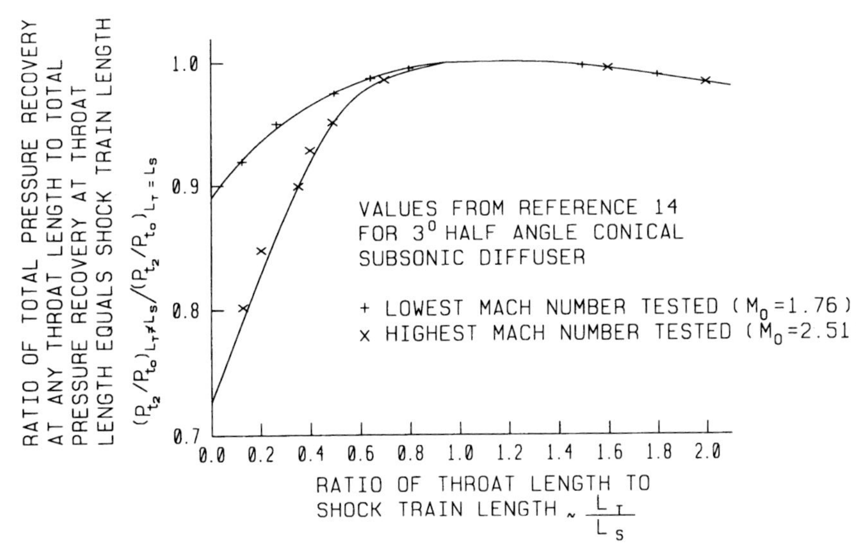

Fig. 3.4 Effect of throat length on inlet total pressure recovery.

In summary, at any design Mach number, the total pressure recovery of a supersonic inlet is sensitive to throat section design. For throat lengths equal to or slightly greater than the shock train length, the recovery is the maximum attainable. For throat lengths shorter than the shock train length, the attainable total pressure recovery decreases with decreasing throat length, and the recovery loss at any throat length is much greater at a higher Mach number than at a lower Mach number. In order to incorporate the proper length throat in any inlet, the designer must know the length of the shock train and the parameters that control shock train length.

It has been shown in Ref. 14 and in other references cited therein that the length of a normal shock in a constant area duct is established primarily by the flow area of the boundary layer as a percentage of the total cross-sectional area of the duct and by the Mach number immediately upstream from the beginning of the shock. Reference 14 recommends the use of an average Mach number at the upstream station, and the M_1 shown in Figs. 3.2 and 3.3 corresponds to the recommended value. The symbol δ_1, in Figs. 3.2 and 3.3, represents the boundary-layer thickness at the beginning of the enclosed throat. However, instead of using boundary-layer thickness, Ref. 14 introduces the boundary-layer displacement thickness δ^* and relates both measured and calculated values of this parameter to the displacement thickness calculated for a 1/7–power velocity profile in a circular duct. This latter thickness has been termed δ^*_{max} and is considered to be the maximum value that will be encountered in practical inlet configurations. The value of δ^*_{max} increases with increasing values of the average Mach number upstream from the beginning of the shock. Data presented in Ref. 14 have been reproduced both directly and in extrapolated form in Fig. 3.5 to illustrate the variation of shock train length with upstream values of average

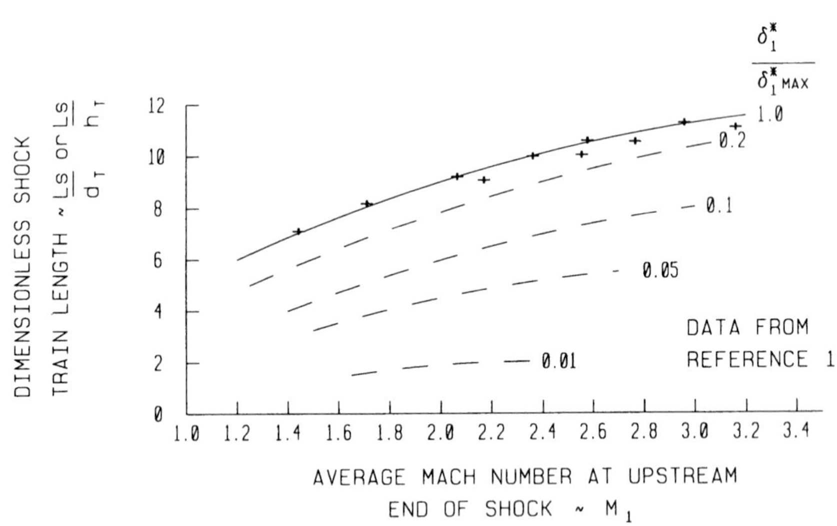

Fig. 3.5 Variation of shock train length with upstream values of average Mach number over a range of values of a boundary-layer thickness parameter.

Mach number M_1 over a range of values of the boundary-layer thickness ratio parameter, $\delta^*_1/\delta^*_{max}$. Shock train length Ls has been made dimensionless by using duct diameter d_T in the case of a circular throat, or channel height h_T in the case of an annular or rectangular throat, as a reference. Although Ref. 14 assigns the number two to the station upstream from the shock, the number one is used herein to maintain consistency in the nomenclature. Generally, shock length increases with upstream Mach number at any value of the boundary-layer thickness ratio parameter, but the rate of increase declines as the boundary layer becomes thinner. Thus, for inviscid flow the shock length is a constant value of zero for any value of the upstream Mach number. At the other extreme, where δ^*_1 equals δ^*_{max}, the shock train almost doubles in length as the upstream Mach number increases from about 1.2 to 3.4.

It is postulated that typical fixed-geometry all-external supersonic compression inlets designed in accordance with the conservative procedures being described throughout this text will encounter a combination of parameters that fall within the limits, or boundaries, defined by Fig. 3.5. Although boundary layers are seldom measured directly in normal wind tunnel tests of inlet model configurations, Schlieren, or shadowgraph, pictures of the external compression flowfield can show any significant main flowfield changes traceable to growth of the boundary layer. For example, if the boundary layer becomes thick enough to increase the effective compression surface coordinates and angles at the design Mach number, the leading-edge wedge or cone shock will lie upstream from the cowl lip. This condition signifies a radical departure from the inviscid design procedure and usually portends the development of a relatively long shock train. Generally, for typical inlets designed at Mach 2.0 and below, Schlieren pictures show little impact of viscous effects on the potential flowfield, and tests with zero throat length confirm the attainment of the maximum total pressure recovery expected. As the design Mach number increases, the boundary layer has an increasingly adverse impact on the potential flowfield, and shock length would be expected to increase with increasing design Mach number as a result of both the increased boundary-layer thickness and the increased value of M_1. A reasonable approximation of dimensionless shock train length with upstream Mach number M_1 for a family of typical conservatively designed supersonic diffusers as defined in Chapter 2 is given in Fig. 3.6 as a single solid line. The values of the design Mach number M_0, which correspond to selected values of the upstream Mach number, are also identified on the lower scale. If the supersonic diffusers were designed with more or less compression than the conservative configurations, then the M_0 scale would have to be shifted to the left or right to achieve the proper correspondence with the M_1 scale, but the single dimensionless shock length curve would probably no longer apply. Because of changes in boundary-layer thickness with variations in the amount of compression, the dimensionless shock lengths would be longer or shorter than the conservative value. As noted in the discussion of Fig. 3.4, it is always safer from a performance standpoint to assume a high value of shock train length and design the throat section accordingly, and the penalty

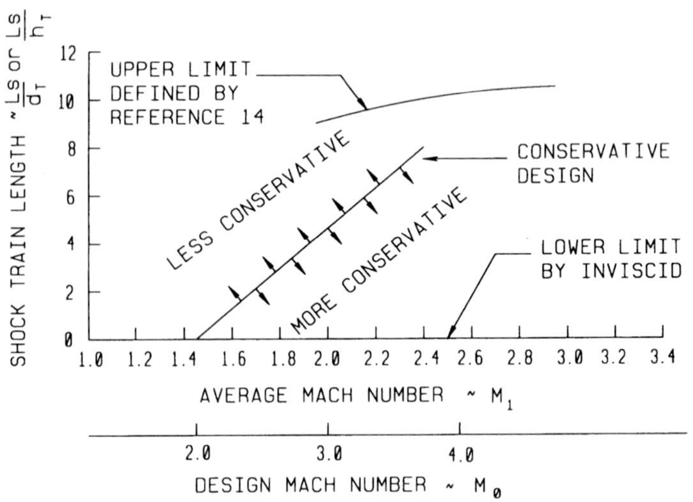

Fig. 3.6 Approximate variation of shock train length with upstream Mach number for typical fixed-geometry all-external supersonic compression inlets.

for too short a throat length is much more severe at a high design Mach number (e.g., 4.0) than at a low design Mach number (e.g., 2.0).

3.3 Design Procedures

In a typical fixed-geometry, all-external supersonic compression inlet, the end of the supersonic diffuser is defined by a cowl lip, a compression surface inclined at an angle θ to the freestream flow direction, and a passage or channel height h, as sketched in Fig. 3.7. The design of the subsonic diffuser begins with these initial boundary conditions coupled with the ever-present requirement to minimize external cowl drag. This latter requirement usually dictates that the inner and outer cowl surfaces be turned, in as short an axial distance as practical, from an inclined angle of θ deg to 0 deg. To avert any internal contraction, the compression surface, the inner body, must also be turned at the same rate as the inner cowl surface. It has been determined in scores of experiments that the rate of turn of the inner body is the key to control of the flow turning process and that a minimum radius of curvature equal to four channel heights provides the shortest practical axial distance in which to complete the turn. If the cowl is flow-aligned, the inner surface coordinates are calculated to maintain the desired length ℓ of the constant-area throat, as determined from Fig. 3.6. For zero throat length the inner cowl surface coordinates are calculated to provide a constantly increasing channel area with increasing distance downstream from the throat plane. If the cowl is misaligned, internal contraction is usually used, and the inner cowl contours must be calculated to provide a decreasing channel area with increasing distance downstream

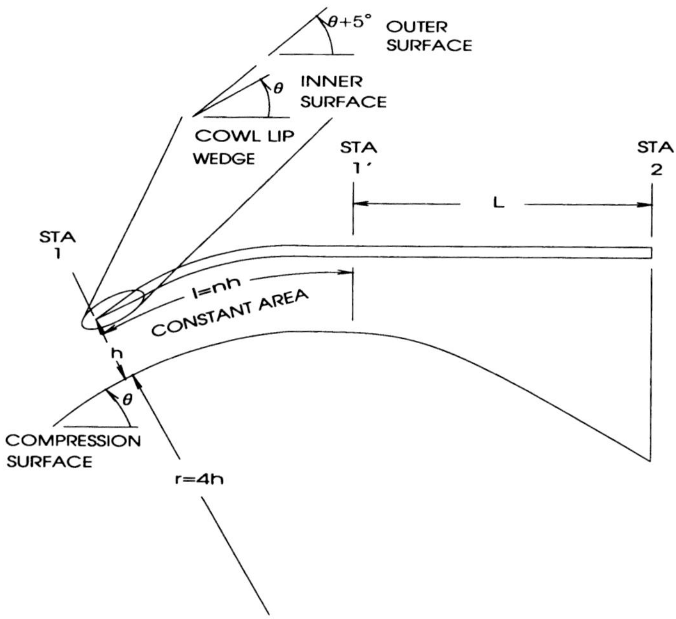

Fig. 3.7 Subsonic diffuser and throat section design parameters.

from the cowl lip plane. For this case the self-starting area ratio restrictions discussed previously determine the minimum area of the channel. Whether the inner cowl surface is aligned or misaligned, the outer cowl surface begins at the cowl lip as a wedge, which, for structural integrity, is usually no less than 5 deg. (A 3-deg included angle has been successfully used on some wind-tunnel models.) Details of the cowl lip wedge are shown in enlarged scale in Fig. 3.7. Downstream from the initial cowl lip wedge, cowl thickness is normally a constant value dictated by structural requirements.

At the end of the throat section, the subsonic flow is diffused in a diverging channel that can have any conceivable geometric shape. The ratio of the exit area A_2 to the entrance area $A_{1'}$ of the diffuser is determined from the equation of continuity written between the two stations:

$$\frac{A_2}{A_{1'}} = \frac{Pt_{1'}\left(\frac{P}{Pt}\mathring{m}\right)_{1'}}{Pt_2\left(\frac{P}{Pt}\mathring{m}\right)_2} \tag{3.1}$$

where

$$\frac{Pt_2}{Pt_{1'}} \sim 0.97$$

for an equivalent conical half-angle on the order of 3–5 deg,

$$\left(\frac{P}{Pt}\,\mathring{m}\right)_{1'}$$

is a function of the Mach number $M_{1'}$ behind the normal shock, and

$$\left(\frac{P}{Pt}\,\mathring{m}\right)_{2}$$

is a function of the selected exit Mach number, M_2. The physical value of $A_{1'}$ equals A_1.

The values of A_1 and A_2 are used to find the radii of equivalent area circles, and these radii in conjunction with a selected value of conical half-angle are substituted into the equation in Fig. 3.1 to solve for the subsonic diffuser length L.

Inasmuch as the subsonic diffuser operates in concert with a supersonic diffuser and throat section to form a complete inlet, it is instructive to examine in detail a complete entity as shaped by differing Mach number requirements. The following paragraphs discuss the integration of a family of conservatively constrained supersonic diffusers, designed in accordance with the procedures presented in Chapter 2, with subsonic diffuser/throat sections, designed according to the procedures discussed earlier. Three examples have been selected to illustrate configurational variances produced by a low, an intermediate, and a high design Mach number in the operational range between Mach 2.0 and 4.0. Figure 3.8 shows a Mach 2.0 design inlet (top), a Mach 3.0 design inlet (center), and a Mach 4.0 design inlet (bottom) drawn to scale. All three inlets compress the flow from the freestream design value to a final subsonic Mach number M_2 of 0.3. All three inlets are two dimensional with a constant width of one unit. The cowl height of each inlet is also one unit, so that the capture area of each inlet is one square unit, and changes in channel height in the axial direction represent changes in cross-sectional area. All three supersonic diffusers are semi-isentropic surfaces with an initial wedge angle of 5 deg and a final compression surface angle that is 11 deg below the turning limit given on Fig. 2.13 (except the Mach 2.0 design, which is about 9 deg below the limit). All three isentropic compression fans focus on the 5-deg wedge shock at each design Mach number at the height of one unit, and the cowl lip for each inlet is located at the focal point. It is evident that the supersonic diffusers increase in length with increasing design Mach number and that the channel height between cowl lip and compression surface decreases. Using the procedure described in connection with Fig. 3.7, each inner body is turned in a radius of curvature equal to four channel heights.

For the Mach 2.0 design supersonic diffuser, the Mach number M_1 at the end of the compression is 1.47. For that value of M_1 Fig. 3.6 indicates an approximate shock train length of zero. Hence, the throat for the Mach 2.0 inlet consists of a plane that contains a single discontinuity with a

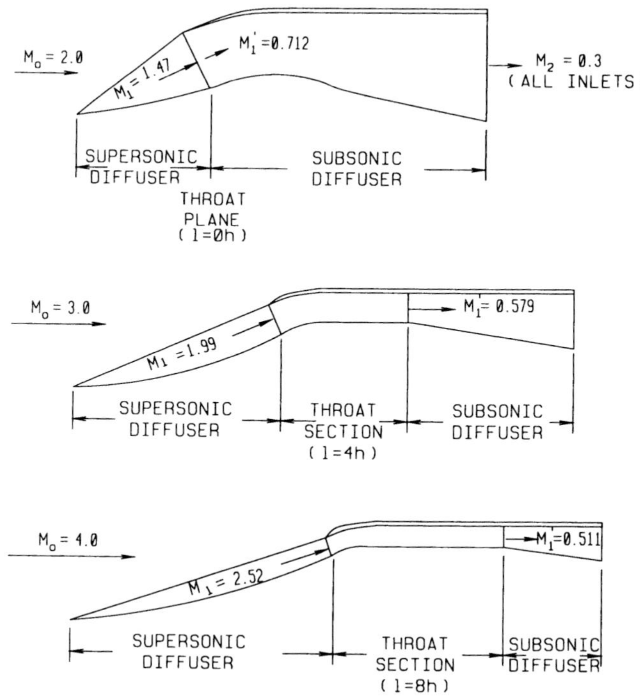

Fig. 3.8 Comparative sketches of a family of inlets designed in accordance with conservative practices.

supersonic stream at $M = 1.47$ on one side and a subsonic stream at $M = 0.712$ on the other side. The inner cowl surface is turned from an initial angle of 15 deg to a final angle of 0 deg with coordinates calculated to provide a steadily increasing cross-sectional channel area. Near the initial point of 0-deg slope for the inner cowl surface, the inner body begins to diverge from the horizontal. The inner-body coordinates are formed by the arc of the circle with radius equal to four channel heights and a line tangent to the arc on the downstream side. This latter line connects the arc of the circle with the inner-body side of the exit plane rectangle, whose area is determined from the continuity [Eq. (3.1)]. The length of the subsonic diffuser is determined by an equivalent conical half-angle equal to 3 deg.

The Mach 3.0 design supersonic diffuser compresses the flow to a Mach number M_1 of 1.99, which requires a throat length of four channel heights to contain the normal shock train, as shown in Fig. 3.6. The throat section is formed on the inner-body side by the arc of the circle with a radius equal to four channel heights and a straight line tangent to the arc at 0-deg slope and on the cowl side by an arc of a circle and a tangent line with coordinates established by the constant channel height requirement. At the end of the throat section, the subsonic diffuser is constructed by an extension of the flat cowl surface and an inclined inner-body surface that connects the end of the throat with the exit plane rectangle. The slope of the inner body is set by the length of the subsonic diffuser as determined by the 3-deg equivalent conical half-angle.

At Mach 4.0, the flow is compressed supersonically to a final Mach number M_1 equal to 2.52, which corresponds to a shock train length on the order of eight channel heights in Fig. 3.6. About eight lengths of constant-area throat are required to contain the shock train. The throat section and the subsonic diffuser for this inlet are constructed in the same manner as described previously for the Mach 3.0 inlet.

Comparisons of the three inlets indicate that the subsonic diffuser is the predominant section at the low Mach number, where the throat section is zero length, but the throat section grows while the subsonic diffuser shrinks with increasing Mach number. Experiments confirm that the relative size of each component reflects a corresponding contribution to overall performance. Inlet performance is more sensitive to subsonic diffuser design at low Mach numbers and to throat section length at high Mach numbers. At Mach numbers where performance is relatively insensitive to subsonic diffuser design, it is customary to increase the equivalent conical half-angle above the "idealistic" value of 3 deg (up to 8 deg has been tested in model scale) and decrease both the length and weight of the inlet. At high design Mach numbers (e.g., 4.0) additional decreases in length and weight can be achieved by dumping the flow from a Mach number such as 0.5 to a lower Mach number of 0.3 or even 0.2. Dumping occurs with an abrupt change (step) in area. The total pressure recovery of the process is decreased by the loss in velocity head. At Mach numbers where the throat section is equal in length or longer than the subsonic diffuser, peak total pressure recovery demands a throat length equal to or slightly longer than shock train length. To ensure a high level of total pressure recovery, albeit somewhat lower than the peak or maximum value, the throat length should be one or two channel heights longer than the anticipated shock train length. If volume constraints preclude the incorporation of an adequate throat length in a particular inlet design, methods to minimize boundary-layer thickness and thereby reduce the shock/boundary-layer interaction length must be explored. (Chapter 4 will discuss boundary-layer control techniques.)

The contributions of the subsonic diffusers and throat sections to the overall total pressure recovery of the three design Mach number inlets of Fig. 3.8 are illustrated in Fig. 3.9, where three different levels of performance are shown. The uppermost curve connects the total pressure recoveries of the three conservatively constrained supersonic diffusers and is a repeat of the middle curve of the five shown in Fig. 2.15. If the flow were inviscid, there would be no losses and no significant flow distortion through the throat section, so that the subsonic diffuser could achieve the high recovery level demonstrated with the diffuser alone. The overall total pressure recovery of the inlet would be the product of the uppermost curve for the supersonic diffuser and a high constant value (0.97 used herein) for the subsonic diffuser. This product is represented by the middle curve in Fig. 3.9. Because of viscosity, the normal shocks interact with the bound-

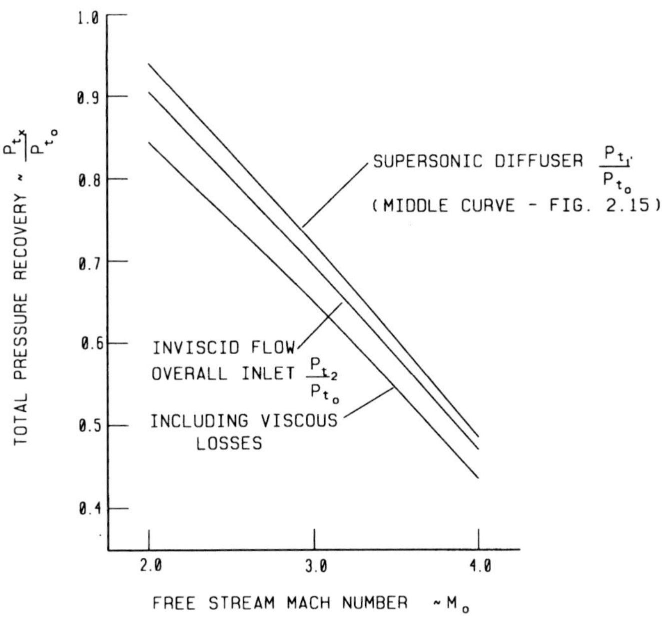

Fig. 3.9 Estimated performance of a family of inlets designed in accordance with conservative practices.

ary layers along the compression surfaces to increase boundary-layer kinetic energy losses and produce slip planes (vortex sheets) with total pressure losses outside the boundary layer in the mainstream. All of the viscous losses have been lumped together to yield the bottom curve in Fig. 3.9. Rather than being treated separately, the viscous losses have been combined with the inviscid performance of the subsonic diffuser and throat section to form a constant, which has a value on the order of 0.90 based on empirical techniques used in the analysis of performance data gathered with many different inlets in scores of wind-tunnel tests. In brief, the inviscid total pressure recovery of a supersonic diffuser (represented here by the upper curve in Fig. 3.9) multiplied by a constant 0.90 yields the overall total pressure recovery of a complete inlet at the design Mach number of the inlet (represented here by the bottom curve in Fig. 3.9). The implication in the above statement is that the throat length and the subsonic diffuser are each designed for peak performance.

Because of viscosity, the flow profile at the entrance to a subsonic diffuser can be distorted. If the distortion persists through the subsonic diffuser with a pattern that interferes with proper fuel distribution for efficient combustion, then flow-straightening devices, such as flow dividers, splitter plates, turning vanes, screens, or aerodynamic grids, should be installed. Even if the entrance flow is free from distortion, volume constraints, which limit throat sections or subsonic diffusers to lengths less than optimum, can cause distortion of the exit flow profile and require the integration of a flow-straightening device in the design procedure. Every flow-straight-

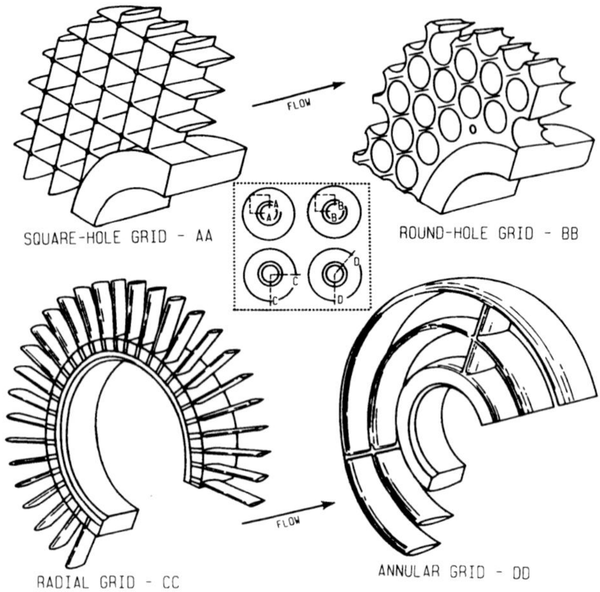

Fig. 3.10 Sections of aerodynamic grids.

ening device produces a drag force that penalizes the total pressure recovery. However, many of the devices can improve the structural integrity of the inlet and actually reduce overall weight. Generally, flow dividers and splitter plates are used for structural purposes. These are flat plates mounted in the vertical or horizontal plane between the walls in the subsonic diffuser and sometimes into the throat section. The plates may be oriented at an angle of attack to the upstream flow so that a potential downstream region of separation may be energized. Turning vanes may be used to impart rotation to the incoming air in the same manner that guide vanes turn the air into the first stage rotor of a compressor. Also, turning vanes have been used to help the flow navigate a diffusing turn without separation. Screens of various solidity have been used extensively in wind tunnels at very low velocity (M less than 0.1), but the total pressure loss penalty incurred by screens increases rapidly with increasing Mach number and is unacceptable in almost all practical inlet applications. The aerodynamic grid is essentially a streamlined screen; i.e., instead of a blunt circular wire forming the boundary between holes in the screen, a streamlined forebody and afterbody attached to the wire provide a low drag profile. The screen and the grid both operate by choking the higher energy mainstream flow and forcing the lower-energy flow to accelerate, so that a stream balance is effected downstream. A properly designed aerogrid can inflict as little as one count in total pressure recovery at the design operating condition. Sections of several kinds of aerodynamic grids are shown in Fig. 3.10, and details of grid designs are discussed in Ref. 15.

4
VISCOUS EFFECTS

4.1 Flow Process

Most aspects of the flow through and around a ramjet engine can be determined with acceptable accuracy by means of the one- or two-dimensional isentropic expressions presented in the foregoing chapters. Deviations from isentropic flow processes are accounted for by the shock-wave equations and by performance parameters such as nozzle efficiency n_N, the nozzle mass flow coefficient C_{N_M}, and the subsonic diffuser total pressure recovery constant. Each deviation from the idealized process penalizes engine thrust capability. In particular, skin friction on all engine surfaces produces an external drag force represented by the expression $\int_0^6 \tau_A \, \mathrm{d}A_w$ in Eq. (1.9) and internal drag forces that appear as total pressure recovery losses across each engine component. Skin friction is caused by viscosity, the property whereby any real fluid adheres to the surface over which it flows.

The effect of viscosity on the fluid layers in proximity to a solid surface is graphically illustrated in Fig. 4.1. The solid surface is an infinitely thin flat plate coincident with the x axis, and the undisturbed stream velocity far from the plate is denoted U. At the plate surface (i.e., $y = 0$), the condition of no slip requires that the local velocity u equal zero. Between the surface and the undisturbed stream, the local velocity varies with distance y from the surface, and a line joining the ends of all of the velocity vectors forms the velocity profile. Newton's friction law states that the shearing stress τ between two adjacent layers of fluid is inversely proportional to the slope of the velocity profile:

$$\tau = \mu \frac{\mathrm{d}u}{\mathrm{d}y} \tag{4.1}$$

where μ is a constant of proportionality, and $\mathrm{d}u/\mathrm{d}y$ is the velocity gradient, $= 1/\tan\theta$.

The constant μ is called the coefficient of viscosity and is large for highly viscous fluids (e.g., lubricating oil) and small for "watery" fluids (e.g., air). For highly viscous fluids the dragging action of the plate will retard the flow at greater distances from the plate than for low-viscosity fluids.

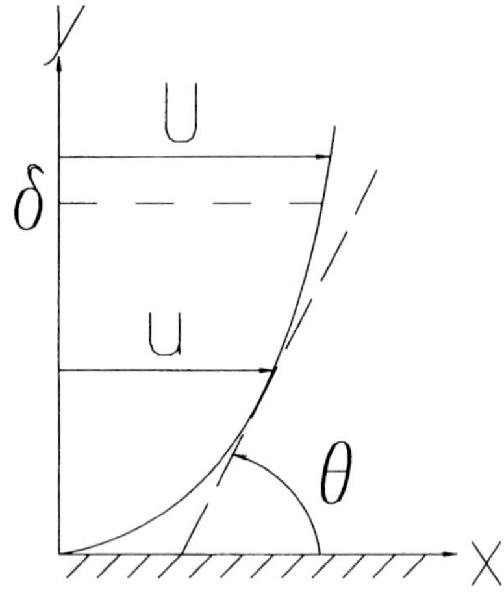

Fig. 4.1 Flow of a real fluid near a wall.

The distance along the y axis at which the disturbance generated by the plate becomes negligible is given by the symbol δ.

Mathematically,

$$u \sim U \qquad \text{at} \qquad y = \delta \tag{4.2}$$

The viscosity of air at 59°F and 1 atm is 3.726×10^{-7} slugs/ft/s. This value is invariant with pressure over the range of conditions explored herein, but increases with increasing temperature in accordance with Sutherland's law.

The relationship between the forces produced by viscous action and the pressure and inertia (momentum) forces previously discussed can best be examined by dimensional analysis. A body characterized by a dimension ℓ moves in one direction at a uniform velocity U through a homogeneous fluid with properties of density ρ and viscosity μ. The aerodynamic forces acting on the body are a function of the four variables, or

$$F = f(\rho, \mu, U, \ell) \tag{4.3}$$

The principle of dimensional homogeneity is applied to derive the nature of the fundamental physical equations:

Pressure force:

$$F_P = pA \sim p\ell^2 \tag{4.4}$$

Inertia force:

$$F_I = ma \sim U^2\ell^2 \tag{4.5}$$

Viscous force:

$$F\mu = \mu \frac{du}{dy} A \sim \mu U \ell \tag{4.6}$$

The ratio of the inertia force [Eq. (4.5)] to the viscous force [Eq. (4.6)] yields

$$\frac{\rho U^2 \ell^2}{\mu U \ell} = \frac{\rho U \ell}{\mu} \qquad \text{(Reynolds number)} \tag{4.7}$$

If the Reynolds number *Re* is large, inertia forces predominate, and viscous forces are negligible. At low Reynolds numbers, the viscous and inertia forces are of the same order of magnitude. Low Reynolds numbers would be expected in the flow region near a solid surface, and the extent of this region, as defined by δ in Fig. 4.1, can be determined from the approximate equality of viscous and inertia force magnitudes.

If the velocity profile shown in Fig. 4.1 is approximated by the linear velocity gradient of Fig. 4.2, the shear stress at the wall,

$$\tau_0 = \mu \frac{U}{\delta} \tag{4.8}$$

From expression (4.5), the inertia force per unit volume,

$$f_I = \rho \frac{U^2}{\ell} \tag{4.9}$$

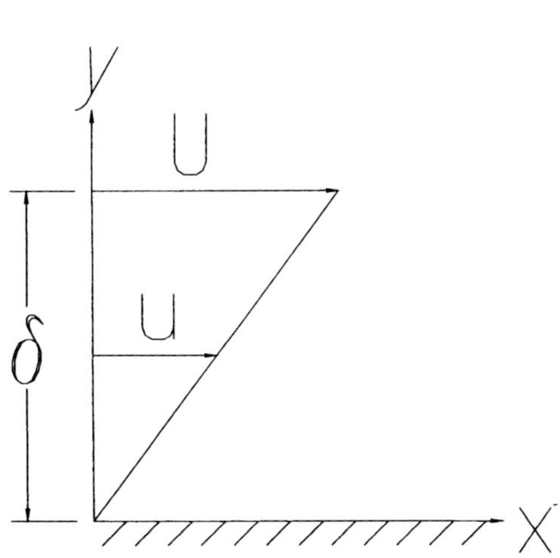

Fig. 4.2 Flow near a wall with linear velocity gradient.

and the force per unit area acting on an element of thickness δ is

$$f_f\delta = \rho U^2 \frac{\delta}{\ell} \tag{4.10}$$

Equating expressions (4.8) and (4.10),

$$\mu \frac{U}{\delta} = \rho U^2 \frac{\delta}{\ell}$$

from which

$$\frac{\delta}{\ell} = \frac{1}{\sqrt{Re}} \tag{4.11}$$

At a flight Mach number of 2.0 at an altitude of 30,000 ft, the Reynolds number 1 ft from the leading edge of any surface is 5.7×10^6. The corresponding thickness δ is 0.0004 ft. Thus, for a representative ramjet engine operating condition, the flow region in which viscous effects are significant is extremely thin in comparison with a characteristic axial length, such as a wing chord or an inlet compression ramp. This thin layer close to a solid surface is called the boundary layer with a thickness measured by δ. In accordance with the boundary-layer theory formulated by Prandtl in 1904, the viscous region can be ignored during the determination of the pressure and inertia forces produced in the main flow region. The subsequent solution to the boundary-layer flow incorporates the influences of the main field forces. Therefore, for many practical aerodynamic problems the potential (main) flow and viscous (boundary-layer) flow can be solved separately with relatively simple equations, and the solutions can be superimposed. However, this simple superposition fails if the boundary-layer thickness δ becomes of the order of the characteristic dimension ℓ in Eq. (4.11).

In the derivation of the boundary-layer thickness equation, a flat plate with no axial static pressure gradient provided the model. If the pressure varies in the direction of flow, an additional term is required in Eq. (4.10) to achieve the balance of forces. If the pressure decreases in the flow direction, as in the expansion through a nozzle, the inertia force is increased, and with no change in the viscous force, the boundary layer is thinner than in the case of zero gradient. If the pressure increases in the flow direction, as in the compression across an inlet, the boundary layer is thicker, and if the pressure gradient becomes sufficiently large to separate the boundary layer, the basic boundary-layer assumption is invalid; i.e., δ is of the order of ℓ, and viscous effects alter the main flowfield. Moreover, for an internal compression process channel height h is a more meaningful characteristic dimension than length ℓ, and boundary-layer thickness δ can be significantly large in comparison with h. In fact, for inlets with substantial internal contraction, the boundary layers growing along opposite walls can overlap and obliterate the potential flowfield.

Of all of the ramjet engine components, the inlet is the most likely to exacerbate any problems associated with viscous flow. In order to enhance the efficiency of the compression process, the inlet designer must have a knowledge of the characteristics and vagaries of the boundary layer and become proficient in the application of boundary-layer control techniques.

4.2 Boundary-Layer Characteristics

The development of a boundary layer along a flat plate moving at a velocity U relative to a real fluid is depicted in Fig. 4.3, where coordinates perpendicular to the plate have been magnified to illustrate some of the salient features of the viscous flow region. The boundary layer grows in thickness along the plate from zero at the leading edge. At inception and for some distance along the plate, the boundary layer is laminar; i.e., the fluid flows in parallel layers, with a velocity profile approximately parabolic in shape. At any point within a laminar boundary layer, the velocity is independent of time. In the region identified as transition in Fig. 4.3, the flow undergoes a transformation from the steady laminar type to a turbulent type where the fluid particles move in random and chaotic fashion with no regard for discrete layers, and the velocity at any point may vary with time in both magnitude and direction. The time average of the velocity in the direction of motion may be represented by a root profile that is fuller than the parabolic laminar profile near the wall and flatter near the outer edge defined by δ. The location of the transition region is a function of the critical Reynolds number at which laminar flow ceases to exist. However, the value of the critical Reynolds number varies with many factors, including surface roughness, initial turbulence of the fluid, and freestream Mach number.

Essential differences between the two viscous flow regimes can be determined by examining the mathematical expressions that define key parameters. The following equations for boundary-layer thickness at any length ℓ on the flat plate, shear stress at the surface, and friction drag coefficient are the result of work conducted during the past century and a half by a multitude of scientists and engineers. The mathematical analyses whereby these equations were derived are too complex for inclusion in this text.

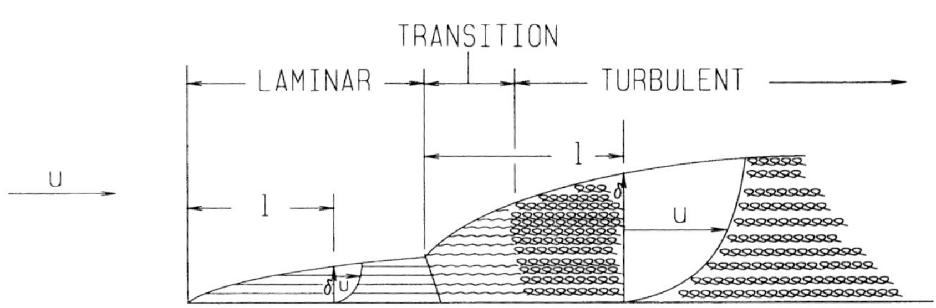

Fig. 4.3 Boundary-layer growth along a flat plate.

Laminar Flow:

The parabolic velocity profile is

$$u = U\frac{y}{\delta}\left(2 - \frac{y}{\delta}\right)$$

Boundary-layer thickness:

$$\delta = \frac{5.48\ell}{\sqrt{Re}} \tag{4.12}$$

Surface shear stress:

$$\tau = 0.730\,\frac{\rho U^2}{2}\sqrt{\frac{1}{Re}} \tag{4.13}$$

Friction drag coefficient:

$$C_f = \frac{1.327}{\sqrt{Re}} \tag{4.14}$$

Turbulent Flow:

The seventh root profile is

$$u = U\,(y/\delta)^{1/7}$$

Boundary-layer thickness:

$$\delta = \frac{0.377\,\ell}{(Re)^{1/5}} \tag{4.15}$$

Surface shear stress:

$$\tau = 0.0587\,\frac{\rho U^2}{2}\left(\frac{1}{Re}\right)^{1/5} \tag{4.16}$$

Friction drag coefficient:

$$C_f = \frac{0.074}{(Re)^{1/5}} \tag{4.17}$$

For a given set of freestream conditions (ρ, U, μ) and a given distance ℓ (from the plate leading edge in the laminar case and from the beginning of transition in the turbulent case), the turbulent boundary layer is thicker, develops more surface shear, and causes a higher friction drag. Thus, long runs of laminar boundary layer on aerodynamic surfaces would be highly desirable, but nature militates against the lower viscous drag flow process in the form of natural transition from laminar to turbulent flow. Despite years of research, there is still great uncertainty concerning the mechanism that triggers transition; and no cogent explanation exists for the wide range of critical Reynolds numbers that have been observed. Values as low as 50,000 and as high as 50 million have been reported in the literature. Generally, the trend, as plotted in Fig. 4.4 for an arbitrary average of observed values, shows a steady increase across the subsonic into the transonic region with a peak near Mach 1.5, and a decrease through the lower supersonic region to a minimum between Mach 3.0 and 4.0, followed by a steady increase across the higher supersonic and hypersonic regions up to Mach 8.0. In the Mach number regime between 2.0 and 4.0, the average values of critical Reynolds number are close to 2×10^6. For the representative ramjet operating condition of Mach 2.0 at 30,000 ft where the Reynolds number per foot is of the order of 6×10^6, a natural transition from laminar to turbulent flow would be completed within one-third of a foot from the leading edge of any surface.

4.3 Boundary-Layer Separation

The boundary-layer equations and the flow examples presented thus far have involved only a flat plate with no pressure gradient, so that the velocity U at the edge of the boundary layer is a constant, and the velocity profiles (parabolic for laminar flow and seventh root for turbulent flow) are invariant with distance along the plate. If a pressure gradient dP/dx is imposed on a surface in the flow direction, the velocity U will change in accordance with Bernoulli's equation as long as the boundary layer adheres to the surface, and the velocity profiles will also change. The flow around a circular cylinder provides the simplest model to illustrate the effect of a pressure gradient. Figure 4.5 shows the flow picture with a) perfect fluid, b) laminar boundary layer, and c) turbulent boundary layer around the

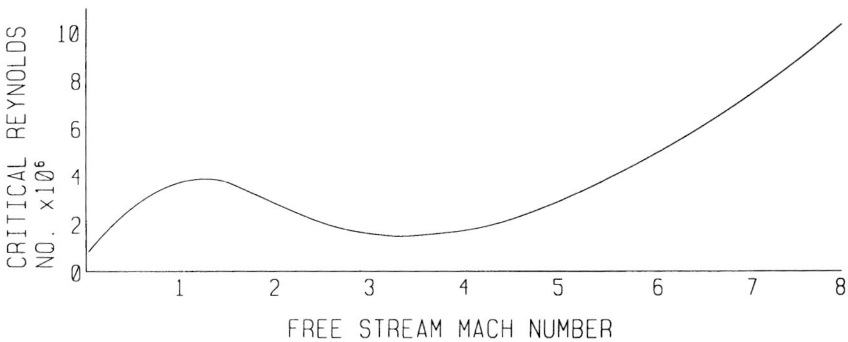

Fig. 4.4 Variation of an arbitrary average of observed critical Reynolds numbers with freestream Mach number.

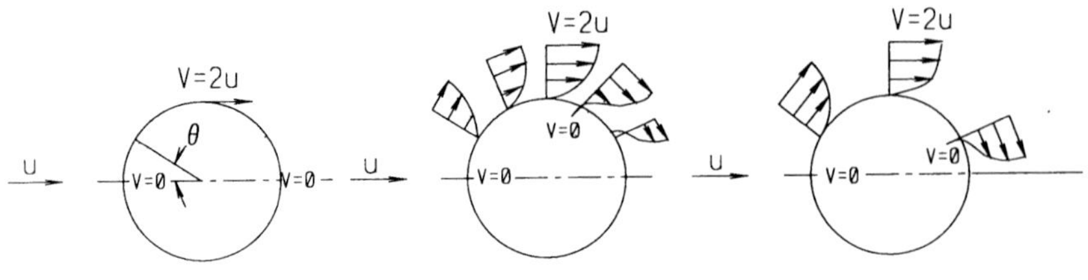

Fig. 4.5 Flow around a circular cylinder: a) perfect fluid; b) laminar boundary layer; c) turbulent boundary layer.

top half of the cylinder. Classical hydrodynamic theory for a perfect fluid shows that the velocity on the surface of the cylinder in case a obeys the expression

$$V_c = 2U \sin\theta \tag{4.18}$$

At $\theta = 0$ deg, the velocity is zero (this is called the forward stagnation point), and the surface pressure is maximum. The velocity increases with the distance around the forward half of the cylinder to a maximum of $2U$ at the top ($\theta = 90$ deg), with a concomitant decrease in pressure from the stagnation value to a minimum at the top. This decreasing pressure gradient over the upstream half of a cylinder is analogous to that encountered in the flow expansion through a nozzle. The reverse process occurs on the downstream half of the cylinder where the pressure gradient increases, as in a compression process, and the velocity decreases to zero at the rear stagnation point ($\theta = 180$ deg). The kinetic energy available to a fluid particle at the top of the cylinder must be just sufficient to transport the particle to the rear stagnation point.

In the case of a real fluid flowing around the cylinder at a Reynolds number below the critical value (Fig. 4.5b), the pressures and velocities outside the boundary layer change in accordance with Bernoulli's law around the upstream half. As the stream velocity increases, the boundary-layer profiles become more full, which is indicative of a higher local Reynolds number and decreased boundary-layer thickness. However, near the surface energy is being converted into heat because of friction, and the fluid particles at $\theta = 90$ deg have less kinetic energy than in case a at the onset of the unfavorable pressure gradient. These surface particles will come to rest before the rear stagnation point is reached and in turn will retard fluid particles a little further from the surface. Eventually, the velocity profile will have a vertical tangent at the wall (the profile in Fig. 4.5b, where $V = 0$), and the main flow separates from the wall. Downstream from this point the flow near the surface actually reverses direction and is usually unstable.

In the case of a real fluid flowing around the cylinder at a Reynolds number above the critical value (Fig. 4.5c), the flow picture is much the same as that in case b, except that the transport of energy across streamlines in turbulent flow produces a much fuller profile and less loss of kinetic energy by the surface particles at the $\theta = 90$ deg position. Thus, the separation point occurs farther around the cylinder and much closer to the rear stagnation point for turbulent than for laminar boundary-layer flow. Unlike the external aerodynamicist, the inlet designer would prefer early transition to turbulent flow in the boundary layer, because turbulent flow is much less susceptible to separation induced by an adverse pressure gradient.

As shown in Chapters 2 and 3, the conversion of kinetic energy to pressure head can be accomplished in a multitude of diffuser shapes that feature converging walls, diverging walls, curved surfaces, inclined surfaces that generate oblique shocks, surfaces that reflect incident oblique shocks, and surfaces upon which normal shocks impinge. The rate at which the area changes in a converging supersonic diffuser or a diverging subsonic diffuser, the rate at which a concave surface is turned, and the strength of a shock that impinges on a diffuser surface determine the level of the adverse pressure gradient that a boundary layer must navigate. The literature abounds with reports of experiments seeking to define the critical level of pressure gradient at which turbulent boundary-layer separation is imminent. One parameter (presented in Ref. 16) relates pressure gradient, boundary-layer profile, and freestream kinetic energy at any point on a compression surface:

$$\frac{\mathrm{d}P}{\mathrm{d}x}\frac{x}{q} \tag{4.19}$$

where $\mathrm{d}P/\mathrm{d}x$ is the adverse pressure gradient; x the distance from the "origin" of the turbulent boundary layer (i.e., the end of transition) and usually the beginning of the supersonic diffuser compression surface; x an indirect measure of boundary-layer profile (i.e., the larger the value, the more susceptible the boundary layer to separation); and q freestream dynamic pressure and a measure of the ability of the freestream to energize the boundary layer and delay separation. This parameter has been applied to the data of Ref. 17, where turbulent boundary-layer separation from a supersonic diffuser was achieved by means of a gradual increase in pressure (a curved surface) and an abrupt increase in pressure (an incident/reflected shock on a flat plate). The two flow situations are shown in Fig. 4.6. In case a the gradient is applied by an isentropic surface and is so gradual that expression (4.19) becomes

$$\frac{\Delta P}{\Delta x}\frac{x}{q} \qquad \text{where} \qquad x = \Delta x$$

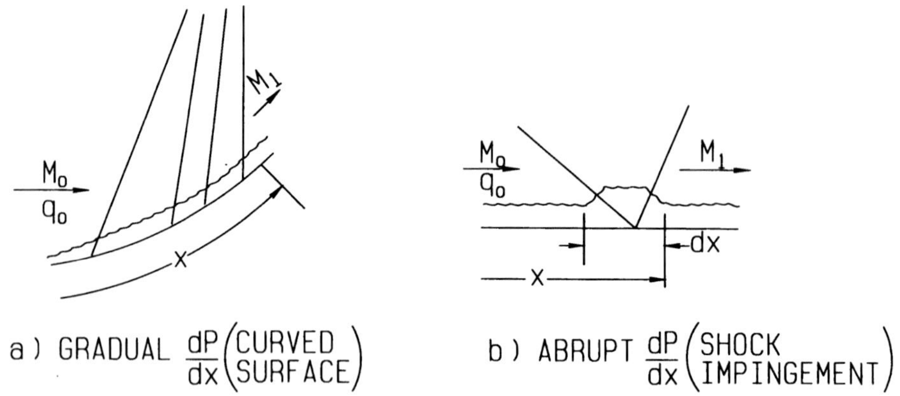

Fig. 4.6 Two examples of application of adverse pressure gradient to a turbulent boundary layer in supersonic flow.

and

$$\frac{\Delta P}{q} = \frac{\dfrac{P_1}{Pt_1} - \dfrac{P_0}{Pt_0}}{\dfrac{q_0}{Pt_0}} \tag{4.20}$$

where $Pt_1 = Pt_0$.

In case b the pressure increase caused by the incident/reflected oblique shock is transmitted forward through the subsonic portion of the boundary layer so that the interaction region dx is greater than the words "abrupt" or "discontinuity" would imply, and dx can be measured with reasonable accuracy:

$$\frac{dP}{dx}\frac{x}{q} = \frac{\dfrac{P_1}{Pt_1}\dfrac{Pt_1}{Pt_0} - \dfrac{P_0}{Pt_0}}{\dfrac{q_0}{Pt_0}}\frac{x}{dx} \tag{4.21}$$

In each flow case the value of the parameter at incipient separation of the turbulent boundary layer was determined from the information presented in Ref. 17 in accordance with the appropriate form of the expression [Eq. (4.20) or (4.21)]. The results are tabulated in Table 4.1.

For case a the critical value of the parameter increases with increasing Mach number, which suggests the confirmation of freestream dynamic pressure as a key in the onset of separation. Although case b data could be obtained only at freestream Mach numbers near 3, the similarity of the critical values of the parameter for the two cases at or near Mach 3 implies

Table 4.1 Critical value of a turbulent boundary layer separation parameter for two compression paths

Case a		Case b	
M_0	$\frac{dP}{dx}\frac{x}{q}$	M_0	$\frac{dP}{dx}\frac{x}{q}$
2.0	.723	3.09	1.58
2.5	1.181	3.06	1.55
3.0	1.548	3.03	1.55

that the separation criteria are the same regardless of the method for applying the adverse pressure gradient.

In Chapter 2 a practical contraction ratio limit was defined and presented as a "running" contraction curve in Fig. 2.14. This limit is roughly equivalent to the isentropic compression of the main flow from freestream Mach number to a final supersonic Mach number equal to one-half the freestream value. Intuitively, the compression limit can be coupled with an adverse pressure gradient level that causes boundary-layer separation. If expression (4.20) is used to determine the critical value of the parameter synonymous with the compression limit, the curve defined by case a can be extended to higher freestream Mach numbers, as shown in Fig. 4.7. Subject to the inaccuracies inherent in the construction of a correlation based on a very limited amount of data, the curve in Fig. 4.7 represents a zone (the critical value plus and minus a band of uncertainty) between a "safe" (attached boundary layer) supersonic diffuser design and a design that will promote boundary-layer separation unless control measures are taken. Thus, Fig. 4.7 provides a yardstick for measuring the probable impact of viscous effects on the supersonic diffuser that evolves from the inviscid design procedures described previously in Chapter 2.

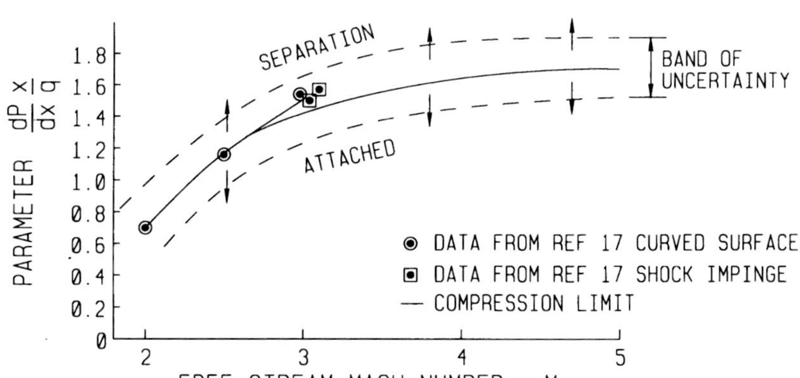

Fig. 4.7 Variation with freestream Mach number of the critical value of a parameter for turbulent boundary layers on supersonic diffuser surfaces.

If the normal shock that is the means of transition from supersonic to subsonic flow in an inlet is positioned upstream from the minimum area section, it is probable that the shock-imposed pressure gradient added to the pressure level existing in the supersonic diffuser will cause the boundary layer to separate. For an all-internal compression configuration, the flow picture will be unstable until the normal shock moves forward of the entrance station (an unstarted inlet). For an all-external compression configuration, the flow picture can also be unstable with the shock moving back and forth to produce a phenomenon called "buzz."

If the normal shock is positioned in the minimum area section of an inlet, the flow picture may be represented by the sketch in Fig. 4.8, where a typical two-dimensional semi-isentropic supersonic diffuser with a flow-aligned cowl is shown. The shock is portrayed as a single discontinuity normal to the local flow direction with one end impinging on a thin laminar layer on the outer (cowl) wall and the other end impinging on a thick turbulent boundary layer on the inner (compression surface) wall. Shock impingement is shown in the form of the Greek letter λ, where the width of the legs is indicative of the interaction zone caused by the pressure rise. In the illustration it is assumed that the pressure rise is less than the critical value required to induce separation of either boundary layer. Downstream from the interaction zone the sum of the boundary-layer thicknesses, $\delta_1 + \delta_2$, is a substantial percentage of the total passage height h; therefore, in the throat region viscous forces are of the same order of magnitude as pressure or inertia forces. If the upstream Mach number M_1 is increased so that the normal shock pressure rise is well above the critical level for boundary-layer separation, the flow picture looks much like the sketch in Fig. 4.9. Boundary-layer separation from the two walls causes a convergence in the main channel flow area so that the subsonic Mach number behind the initial normal shock must accelerate at least to Mach 1 and probably into the supersonic region, where a second normal shock occurs and produces a set of conditions that causes the process to be repeated. The number of shocks in the train increases with increasing strength of the

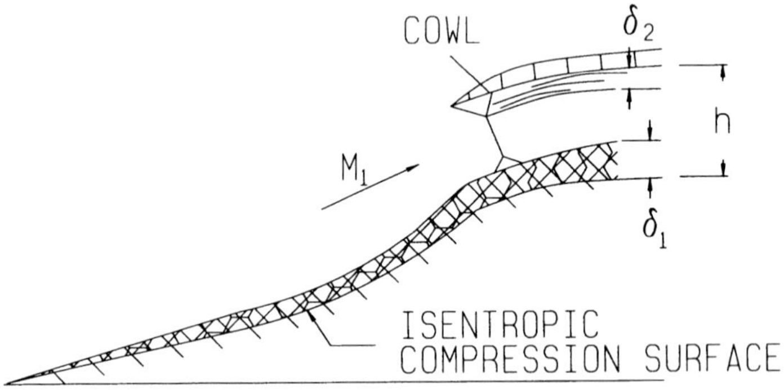

Fig. 4.8 Sketch of low-strength normal shock in an inlet throat with no boundary-layer separation.

original shock, increasing thicknesses of the boundary layers at initial shock impingement, and decreasing height of the throat passage. As shown in the representative inlet drawings of Fig. 3.6, all of the conditions for increasing shock train length go hand in hand with increasing inlet design Mach number.

The preceding discussion and the sketches of Figs. 4.8 and 4.9 constitute a qualitative assessment of normal shock/boundary-layer interaction in the minimum area section of an enclosed channel. Because the shock interacts with multiple boundary layers, because a characteristic length ℓ or x seems irrelevant, and because the interaction length dx is analytically unknown, the parameter $(dP/dx)(x/q)$ probably has no practical significance. Until more specific viscous effects data become available, it seems reasonable to design inlet throat sections in accordance with the empirical information contained in Ref. 14, as described and explained in Chapter 3.

If the normal shock is positioned downstream from the minimum area section, it is probable that the boundary layer will separate under the additive effects of the shock-induced pressure gradient and the pressure gradient imposed by the diverging walls of the subsonic diffuser. Unlike the case where the normal shock is positioned on the upstream side of the throat, the normal shock/subsonic diffuser boundary-layer interaction is usually stable. Also, any boundary-layer separation induced flow distortion that produces combustion problems can be assuaged with one of the flow-straightening devices described in Chapter 3.

If the flow in the subsonic diffuser is shock-free, but the wall boundary layers have already navigated the adverse pressure gradients imposed by a supersonic diffuser and/or a throat section, it seems reasonable to expect that the boundary-layer velocity profiles are approaching the vertical tangent shape indicative of incipient separation, as shown in Fig. 4.5b or 4.5c, where $V = 0$. Although the parameter $(dP/dx)(x/q)$ could conceivably be used to assign a quantitative value to the separation proneness of the boundary layer, it is not known how to determine a realistic value of x in order to correlate data from a variety of configurations. Generally, the

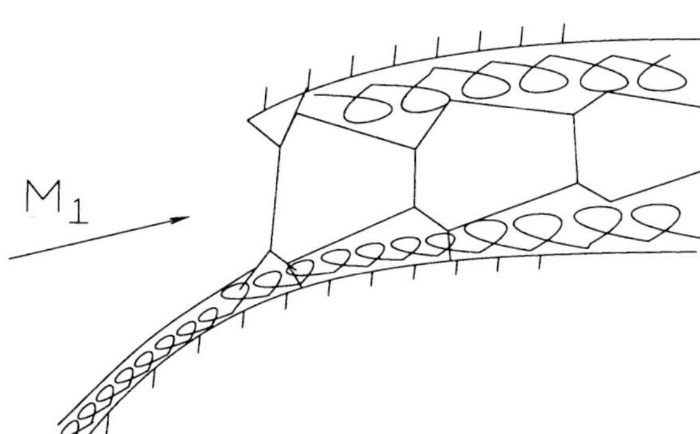

Fig. 4.9 Sketch of high-strength normal shock with boundary-layer separation and development of a shock train.

equivalent conical half-angle rule described in Chapter 3 for designing a subsonic diffuser precludes the imposition of a critical pressure gradient, so that the inviscid design procedure tacitly incorporates viscous effects.

4.4 Boundary-Layer Control

Inasmuch as a turbulent boundary layer can withstand a higher pressure gradient than a laminar boundary layer before separation occurs, the inlet designer welcomes natural transition near the leading edge of any compression surface. For most flight trajectories of practical interest, the operating Reynolds number per foot of surface length is higher than the critical Reynolds number to the extent that natural transition is considered to occur at the leading edge. However, for some trajectories it is possible to have an operating Reynolds number equal to or even lower than the critical value so that laminar boundary layer exists along a considerable length of the compression surface. Schlieren pictures of the flow over an isentropic compression surface (the sketch in Fig. 4.10) show that the laminar boundary layer separates but reattaches to form a bridge across the contoured shape so that the prescribed amount of main flow turning cannot take place, and inlet performance (both total pressure recovery and capture area ratio) is less than desired. Usually, upon reattachment, transition occurs, but the inlet performance has already been compromised. In order to force transition upstream from a curved compression surface, trip devices can be installed on or integral with the initial wedge (or cone), which contains the leading edge of the compression surface. At Mach numbers between 2 and 4, laminar boundary layers have been successfully tripped with either grit or wires, as sketched in Fig. 4.11. However, at Mach numbers of 5 and above, transition could only be effected with three-dimensional trips that resemble gear teeth, as shown in Fig. 4.11. Regardless of trip device geometry, the trip height must approximate the boundary-layer thickness at the trip location in order to achieve transition with minimum performance penalty. If the trip height is much less than the boundary-layer thickness, transition is not triggered. If the trip height is much more than the boundary-layer thickness, shocks generated by the protruding trip device cause total pressure losses that penalize inlet per-

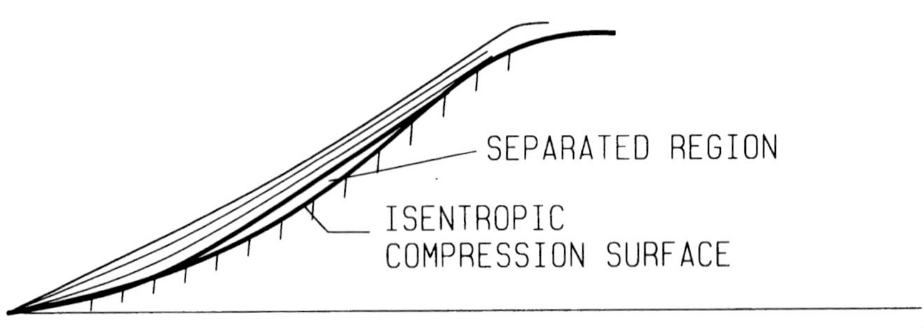

Fig. 4.10 Sketch of a laminar boundary-layer bridging across an isentropic compression surface.

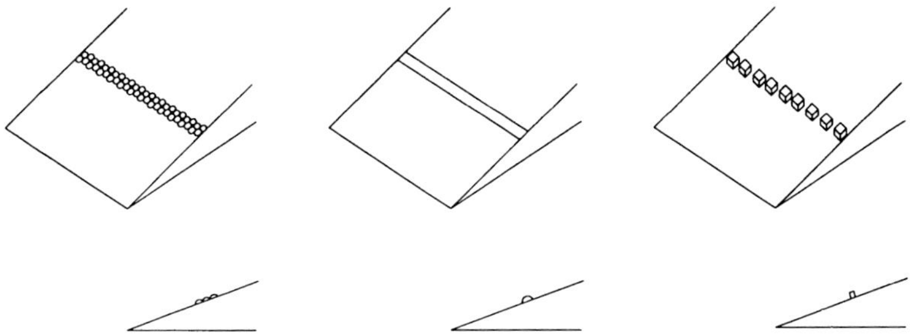

Fig. 4.11 Boundary-layer trip devices: a) grit; b) wire; c) teeth.

formance. In any inlet development program it is prudent to plan tests with a range of trip heights that encompasses boundary-layer thickness estimates for several flight operating conditions, where forced transition is likely to be required.

Whether transition near the leading edge of an inlet compression surface is natural or forced, the mere presence of a turbulent boundary layer does not guarantee immunity from separation. If a supersonic diffuser is designed with a value of the parameter, $(\mathrm{d}P/\mathrm{d}x)(x/q)$, which exceeds the critical level shown in Fig. 4.7, the turbulent boundary layer will separate. A throat length that is too short to contain the complete normal shock train will result in turbulent boundary-layer separation in the subsonic diffuser, as shown in Fig. 3.2. Even with shock-free flow, a subsonic diffuser with too high a divergence angle can cause boundary-layer separation. In each case the obvious solution is to increase length: stretch the axial coordinates of the supersonic diffuser, increase the length of the throat section, and decrease the divergence angle of the subsonic diffuser. However, for many missile applications the inlet length is limited by constraints over which the inlet designer has no control, and the maximum permissible inlet length could violate all of the criteria for attached boundary-layer flow.

One successful technique for overcoming the boundary-layer separation proclivities of a "too short" inlet is boundary-layer bleed, wherein the low energy layer adjacent to the compression surface is diverted through a separate passage. A boundary-layer bleed arrangement is shown in the sketch in Fig. 4.12. The bleed flow is extracted through an entrance, collected in the plenum, and exhausted through a nozzle whose centerline parallels as closely as possible the centerline of the ramjet engine exit nozzle. For the arrangement shown in the sketch, the nozzle is located in the vehicle boundary-layer diverter, which prevents the ingestion of vehicle boundary layer by the inlet. For axisymmetric podded engines compression surface bleed would be collected in the centerbody and exhausted through hollow struts that also serve as a structural bond between the centerbody and cowl. In the latter case the strut-mounted nozzle centerlines would be inclined at some angle with the engine centerline, and some of the axial momentum of the bleed flow would be lost. Generally, the size and shape

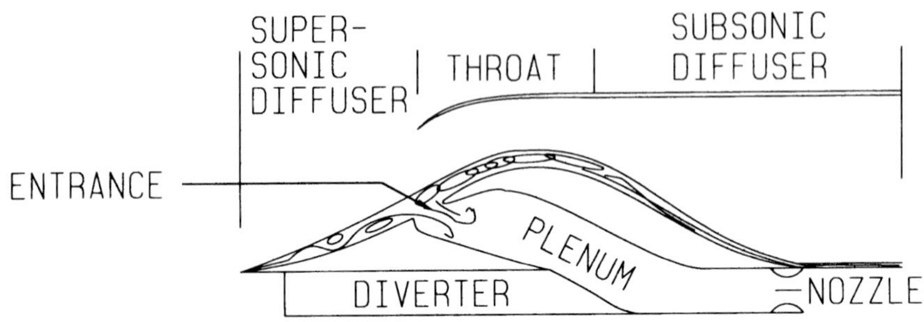

Fig. 4.12 Boundary-layer bleed arrangement.

of the bleed constituents depend on the size and shape of the volume available for bleed disposal, and each bleed arrangement is tailored to fit a specific installation. Furthermore, not all of the constituents shown in the sketch are present in every bleed arrangement. For example, bleed extracted from an inlet side wall or cowl would simply be dumped overboard into the ambient stream. The bleed holes or slots could be drilled or milled at an angle to recover some of the axial momentum and reduce the bleed drag.

Some typical bleed entrances are shown in the sketches in Fig. 4.13. Flush holes in the staggered pattern shown in Fig. 4.13a are usually favored for supersonic diffuser bleed. Almost any size drill can be used, and, during wind-tunnel tests of inlet models, holes can be plugged (with dental plaster, tunnel wax, tape, etc.) and unplugged in an infinite variety of patterns to vary both the amount and location of the bleed. The static pressure of the potential flow at the location of a flush hole is the driving pressure of the bleed configuration. Because the axis of the flush hole is at right angles with the flow and the driving pressure is so low, the flow coefficient of a flush hole is only of the order of 0.50. The flush slot depicted in Fig. 4.13b is most often used in or near the throat region as a shock trap, i.e., to catch the oblique shock generated by a misaligned cowl, prevent the formation of a normal shock train, or stabilize the normal shock in a subcritical position to avoid buzz. The driving pressure of the bleed varies with shock strength and position relative to the slot station, and, generally, the slot has a higher flow coefficient than the holes. However, slot geometry is relatively inflexible in terms of a convenient variation of flow area and location during a wind-tunnel test. The ram slot (Fig. 4.13c) is used to

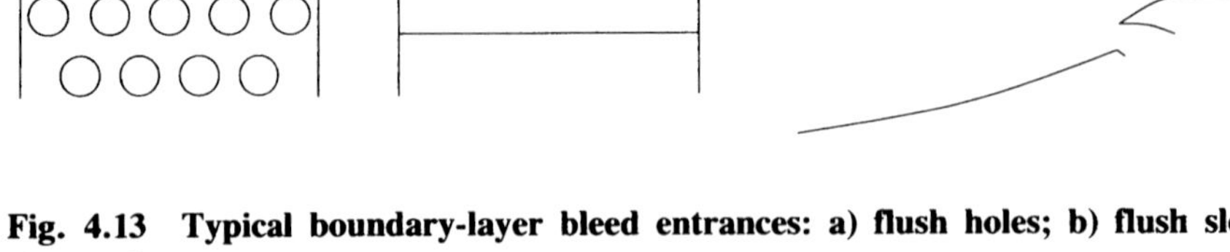

Fig. 4.13 Typical boundary-layer bleed entrances: a) flush holes; b) flush slot; c) ram slot.

skim off the low-energy boundary layer at the end of the supersonic diffuser and minimize viscous effects on the performance of the throat and subsonic diffuser. The driving pressure of this entrance approaches the stagnation pressure of the potential flow at the slot station, and the flow coefficient approaches unity. This type of entrance has two disadvantages: 1) the geometrical inflexibility associated with slots in general; and 2) the potential for ingesting large quantities of high-energy flow at off-design inlet operation, where the fixed height of the slot is much greater than the compression surface boundary-layer thickness. High-energy flow diverted through the bleed passage is unavailable for thrust generation, and the bleed drag is needlessly increased. Determination of bleed drag, comparison with other types of drag, and the trades between drag and thrust contributions for boundary-layer control by bleed are discussed in Chapter 6.

One advantage ascribed earlier to boundary-layer control by bleed is the ability to achieve acceptable performance with an inlet that is shorter than a safe (no boundary-layer separation in any section) length. The safe length can be determined with reasonable accuracy by means of the empirical procedures previously presented, but there is no known correlation between inlet length decrease and bleed flow rate. The problem has been addressed experimentally but with only one variable (i.e., bleed flow rate), and the results for a few selected inlets are given in Fig. 4.14. The inlets have nothing in common except that the bleed flow rates were varied during wind-tunnel test programs. The two inlets with flush holes were tested over a range of bleed flow rates from 0 to 5% of the captured mass flow. The NASA inlets[18] could not be operated at zero bleed, and, in fact, the minimum bleed for inlet start was of the order of 6% through a flush slot. Combinations of flush slots for starting bleed and ram slots for up to full boundary-layer removal were used to achieve bleed flow rates as high as 26% of the captured mass flow. The two-dimensional fixed-geometry inlet[19] could be operated with all bleed holes closed, but the measured total pressure recovery is less than the normal shock value at the freestream Mach number of 2.38. A bleed flow rate of only 1% increases the total pressure recovery of the inlet by more than 20%. Inlet performance continues to increase with additional 1% increments in bleed, but the performance gains are progressively less. Although the axisymmetric variable-geometry inlet[20] could be started without bleed and could achieve a total pressure recovery much higher than the normal shock value of 0.511 at a freestream Mach number of 2.47, performance gains can be realized with the application of bleed in increments of 1%, as shown in Fig. 4.14. As the bleed flow rates reach the 5% level for both of the inlets with flush holes, the total pressure recovery curves approach the same slope, which is very similar to the slope of the recovery curves for the inlets with flush and ram slots. Because the results presented in the literature have not embraced a systematic exploration of all of the independent variables involved in the application of bleed to an inlet, it is inadvisable to reach any conclusions regarding the significance of the common slope of the inlet pressure recovery vs bleed flow rate curves. The curves given in Fig. 4.14 are intended only to portray representative maps of inlet performance that can be achieved

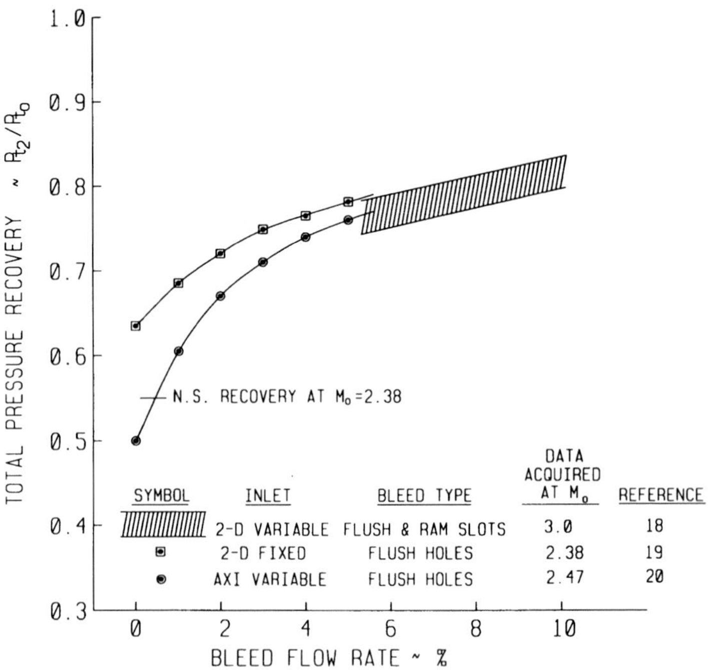

Fig. 4.14 Effect of bleed on performance of three unrelated inlets.

with various amounts of boundary-layer bleed and three bleed entrance geometries. Such a map can be developed for any specific inlet over a range of Mach numbers during initial exploratory wind-tunnel tests, if a sufficient number of bleed geometry variables (entrance and exit) are incorporated into the inlet model and a sufficient number of test runs are completed. These maps are not end items, but provide necessary data to conduct performance trades whereby a best compromise bleed (entrance and exit size) configuration can be selected. The estimated performance for the selected configuration is usually confirmed in subsequent second entry wind-tunnel documentation tests.

Another demonstrated technique for controlling boundary layer is inlet wall cooling. Very limited data were obtained in the hypersonic flow regime by the United Aircraft Research Laboratories during wind-tunnel tests in the early 1960s of an axisymmetric variable-geometry inlet model. Inlet total pressure recovery measured at Mach 8.0 with a cooled wall model was comparable to that of an uncooled model with 8% bleed. The penalties associated with wall cooling in flight are unknown. The most logical choice for a cooling medium would be engine fuel, which could be routed from the tank through passages in the inlet compression surfaces to the injectors, and the heated fuel would enhance the combustion process. However, a host of variables still need to be explored before wall cooling can be considered as a practical alternative to boundary-layer bleed in a flight inlet.

5
OPERATIONAL CHARACTERISTICS

5.1 On-Design Conditions

The definition of on-design operation for a typical ramjet inlet contains three basic conditions:

1) operation at the freestream Mach number for which the compression surfaces were designed;
2) operation with the supersonic compression surface leading-edge shock and the compression fan focal point on the cowl lip; and
3) operation with the terminal normal shock located in the throat section of the inlet, an operating condition called "critical."

A typical ramjet inlet is defined as a fixed-geometry, external compression, semi-isentropic surface member of the two-dimensional family, and a sketch of the on-design flow picture for the typical inlet is shown in Fig. 5.1.

The definition of on-design operation for inlets other than the typical one selected herein can differ markedly from the conditions described earlier and illustrated in Fig. 5.1. For example, the inlet designer may choose to locate the cowl lip downstream from the wedge shock and focal point in order to improve angle-of-attack performance. Many mixed compression inlets are designed with the cowl lip at a lower height than the focal point and upstream from the wedge shock so that the shock and the isentropic compression waves are reflected from the inner cowl surface to increase the internal compression. Most inlets for turbojet application operate with the terminal normal shock upstream from the cowl lip to ensure minimum possible distortion of the flow delivered to the compressor. Thus, the definition of on-design operation for a particular inlet can comprise any one of a diverse combination of conditions that can be controlled by the inlet designer.

Inasmuch as an inlet must work in concert with a combustion chamber and exit nozzle to produce thrust adequate for sustained flight of a vehicle/engine combination over a range of Mach numbers and altitudes, the selection of an inlet design Mach number is predicated upon engine capability to satisfy thrust levels required by cruise, or acceleration, or terminal

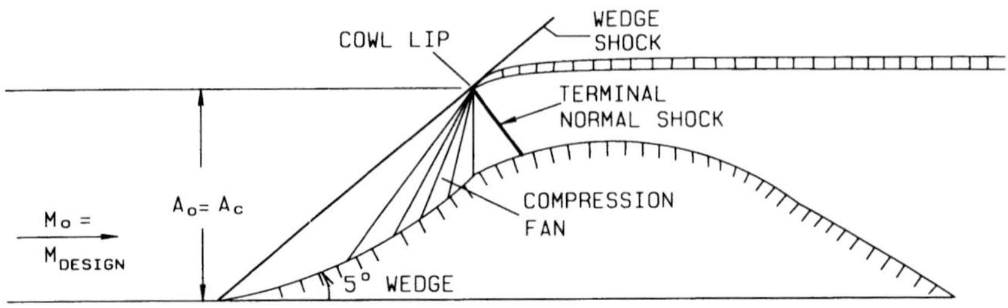

Fig. 5.1 Sketch of typical ramjet inlet operating on design.

maneuver, or all three in some sequence. As described in Chapter 2, the logical choice of an inlet design Mach number for a cruise mission, where the missile is boosted on rocket power to the cruise Mach number, is obvious. However, if the ramjet engine is required to accelerate the missile from an end of boost Mach number to a higher cruise Mach number the choice of inlet design Mach number is not so obvious. Experience has shown that an inlet design Mach number halfway between the end of boost and cruise values offers a reasonable compromise between the accelerating and cruise thrust requirements. If terminal maneuvering requirements are added to the accelerating and cruise specifications, the inlet design Mach number would probably be shaded toward the higher end of the Mach number range. In many practical applications where a variety of thrust levels must be attained, inlet performance is estimated at three or more design Mach numbers, and the final choice is determined from missile performance results generated by trajectory analysis routines.

Usually all of the engine components are designed to produce maximum net thrust at the inlet design Mach number. The inlet contributes the highest total pressure recovery consistent with full capture and minimum drag for a prescribed geometry. This condition is achieved for the typical fixed-geometry inlet when the cowl lip is located on the wedge shock at the focal point of the compression fan, and the terminal normal shock is positioned in the throat (i.e., the critical position). The combustor contributes the minimum possible total pressure loss with the highest achievable heat release, and these objectives are compatible at stoichiometric fuel-to-air ratio, as discussed in Chapter 1. The nozzle throat area A_5 is sized so that the continuity [Eq. (1.26)] is satisfied with the terminal normal shock standing in the inlet throat at stoichiometric heat release. At this juncture, it is helpful to review the terms that comprise the continuity equation:

$$\frac{Pt_5}{Pt_0} = \frac{\left(\dfrac{P}{Pt}\mathring{m}\right)_0 A_0\left(1 + \dfrac{f}{a}\right)}{\left(\dfrac{P}{Pt}\mathring{m}\right)_5 A_5 C_{N_M}} \sqrt{\frac{Tt_5}{Tt_0}} \tag{1.26}$$

where (Pt_5/Pt_0) is the total pressure recovery across the engine,

$$= \frac{Pt_5}{Pt_4}\frac{Pt_4}{Pt_2}\frac{Pt_2}{Pt_0}$$

where the variable Pt_5 is for the nozzle, Pt_4 for the combustor, and Pt_0 for the inlet, $(Pt_5/Pt_4) = 1.0$ [see Eq. (1.6)], (Pt_4/Pt_2) is nearly constant over a wide range of f/a ratios for a given A_5 (see Fig. 1.8),

$$\left(\frac{P}{Pt}\mathring{m}\right)_0$$

is a function of only the selected freestream Mach number M_0,

$$\left(\frac{P}{Pt}\mathring{m}\right)_5$$

is uniquely defined by $M_5 = 1.0$, $A_0 = A_C$ at full capture, $A_5 C_{N_M}$ is chosen so that Pt_2/Pt_0 is the critical value at stoichiometric f/a, f/a is an independent value, $(1 + f/a) = 1.07$ for $JP - 5$ at stoichiometric, $\sqrt{(Tt_5/Tt_0)}$ is given by Eq. (1.29) and Fig. 1.5 for a specified M_0, altitude, f/a, and combustion efficiency n_C.

As soon as the engine is sized (A_0, A_C, and A_5 known) and the operating conditions are selected (M_0 and altitude known), many of the terms in Eq. (1.26) may be recast as constants:

$$\frac{Pt_4}{Pt_2} = K_1; \qquad \frac{\left(\frac{P}{Pt}\mathring{m}\right)_0}{\left(\frac{P}{Pt}\mathring{m}\right)_5} = K_2; \qquad \frac{A_0}{A_5 C_{N_M}} = K_3$$

and continuity may be written as

$$K_1 \frac{Pt_2}{Pt_0} = K_2 K_3 \left(1 + \frac{f}{a}\right) \sqrt{\frac{Tt_5}{Tt_0}} \tag{5.1}$$

At the selected value of inlet design Mach number, the critical total pressure recovery (the highest value of Pt_2/Pt_0 attainable with $A_0 = A_C$) of the inlet is reached when the heat addition parameter $[(1 + f/a)\sqrt{Tt_5/Tt_0}]$ has a value that is consistent with stoichiometric f/a.

To reduce engine thrust for a steady-state cruise requirement, the f/a is reduced, which in turn decreases the value of the heat addition parameter, and the total pressure recovery of the inlet, Pt_2/Pt_0, must be less than the critical value in order to satisfy Eq. (5.1). Obviously, at reduced heat release, the terminal normal shock must move downstream from the throat section into the subsonic diffuser, where the diverging passage permits the flow to accelerate to higher Mach numbers and correspondingly lower normal shock total pressure recoveries. This condition is called "supercritical."

Under idealized operating conditions, which include standard atmosphere, steady-state flow with no disturbances, and a perfect fuel control, it would be impossible to force the terminal normal shock forward of the throat, or critical, position at the inlet design Mach number. Because of abnormalities and imperfections associated with hot and cold atmospheric conditions, wind gusts, viscous flows, differential thermal growth, ablation of thermal protection material, combustor pressure oscillations, and errors in metering fuel flow, the constants K_1, K_2, and K_3 in Eq. (5.1) can have a range of values, and the solutions to the continuity equation can admit a condition where the terminal normal shock stands upstream from the cowl lip. This condition is called "subcritical." Of course, it is possible to design deliberately for such a condition as is common practice with turbojet inlets. The nozzle throat A_5 is sized so that the continuity [Eq. (1.26)] is satisfied with the terminal normal shock standing in the inlet throat at a heat addition parameter less than the stoichiometric value. When f/a is increased toward the stoichiometric value, Eq. (1.26) can be satisfied only if A_0 becomes less than A_C; i.e., the inlet is spilling air. This situation is accomplished with the terminal normal shock standing upstream from the cowl lip. As long as the shock maintains a stable position, the total pressure recovery of the inlet is about the same as the critical value. Operation at freestream Mach numbers other than the design value can also produce a steady-state, standard day, unperturbed flow condition where continuity is satisfied with the terminal normal shock standing upstream from the inlet cowl lip.

Although the on-design set of conditions is essential for establishing engine thrust capability, in most trajectories of current practical interest the engine operates on-design only for brief periods, and most of the flight is accomplished at Mach numbers either higher or lower than the design value and at terminal normal shock positions on the downstream side of critical. Operation with the terminal normal shock positioned upstream from the critical location is always possible, is a deliberate design feature of an inlet for turbojet application, and is carefully circumvented in the control of an inlet for ramjet application. Of course, the ramjet inlet can be designed to preclude any possibility of subcritical operation during flight; i.e., the exit nozzle throat is sized at a value larger than that for critical total pressure recovery, so that at stoichiometric heat release, the inlet can achieve only about 90% of its critical value capability; thus, a supercritical margin is built into the design.

5.2 Off-Design Conditions

There are eight possible off-design operating modes at 0-deg angle of attack, and they are best portrayed in the tic-tac-toe format with the on-

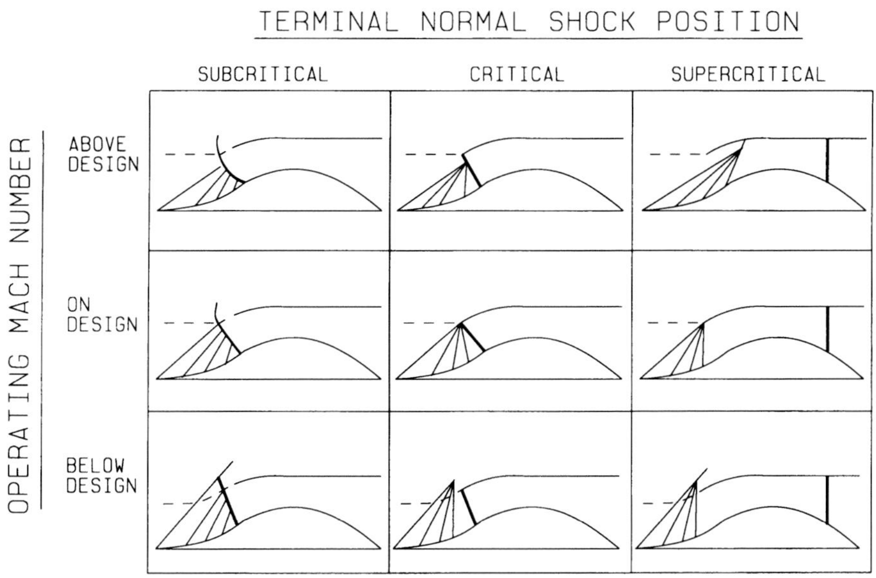

Fig. 5.2 Sketches of nine operating modes of a typical ramjet inlet.

design flow picture occupying the middle square, or box, as shown in Fig. 5.2. The horizontal rows of boxes show the changes in the external shock and compression fan structure as the inlet operates over a Mach number range from an above-design value (upper row) through the design point (middle row) to a below design value (bottom row). The vertical rows of boxes show a representative terminal normal shock position for supercritical operation (right row), critical operation (middle row), and for subcritical operation (left row). In each of the nine sketches the captured stream tube boundary is shown as a dashed line. At the design Mach number of the inlet, the external shock and compression fan structure are unchanged by terminal normal shock movement from the critical position to any supercritical position. Consequently, the captured stream tube boundary is unchanged, and the cross-sectional area of the captured stream tube equals the cross-sectional area of the cowl at all supercritical operating conditions. However, if the terminal normal shock moves forward of the cowl lip in the subcritical position, the external shock and compression fan merge with the normal and produce a bowed, or curved, shock with a series of slip planes or vortex sheets in the subsonic region. The most abrupt change in normal shock direction occurs at the intersection of the normal and the wedge shocks, where the Mach numbers and flow directions upstream from the normal shock exhibit the greatest differences. The curved slip plane downstream from this intersection defines the boundary between the externally compressed flow captured by the inlet and the uncompressed freestream flow that is spilled over the outside of the cowl. In brief, the capture area of the inlet is less than unity.

At Mach numbers above the design value, both the wedge shock and the compression fan focal point move inside the cowl lip so that the captured stream tube consists of an uncompressed freestream flow as well as the externally compressed flow stream. At the critical condition, the wedge shock and compression fan merge with the normal shock and produce curvature with the resultant downstream slip planes. During supercritical operation, the merger of the wedge shock and the compression fan gives birth to a new shock, which is always stronger than the wedge shock, and which can reflect from the cowl wall. The strength of this new shock is a function of many variables, which include the amount of isentropic turning, the initial cowl lip angle, and the delta Mach number above the design value, and has, as a limit, a normal shock at the freestream Mach number. This latter condition is reached when the compression limit (Fig. 2.13) requires the static pressure behind a normal shock to balance the static pressure of the compressed flow on the inner-body side of the vortex sheet. For the typical inlet used herein as a model, the external flow picture is unchanged by terminal normal shock movement from the critical position to any supercritical position during operation at any Mach number above the design value. Consequently, the captured stream tube boundary is unchanged, and the capture area ratio is unity at all supercritical operating conditions. During subcritical operation, the external shock and compression fan merge with the normal to produce a curved shock pattern similar to that observed at the design Mach number, and the inlet spills air over the cowl lip.

At Mach numbers below the design value, both the wedge shock and the compression fan focal point fall outside of the cowl lip, and a streamline traced through the compression field from the freestream to the cowl lip shows that the captured stream tube area is less than the cowl area. The external flow picture is unchanged by movement of the terminal normal shock from the critical position to any supercritical position. During subcritical operation the curved normal shock standing upstream from the cowl lip increases the spillage compared with the critical/supercritical level.

Based on an assessment of the flow picture that prevails at each of the nine possible operating modes of the typical ramjet inlet, it is possible to construct a qualitative diagram of the performance capability associated with variations in the terminal normal shock position and in flight Mach number. Such a diagram is shown in Fig. 5.3, where total pressure recovery is plotted against capture area ratio. At any given Mach number, as the *f/a* is increased from zero to stoichiometric, the terminal normal shock moves from a supercritical position well downstream in the subsonic diffuser to the critical position in the inlet throat. The corresponding total pressure recovery, Pt_2/Pt_0, increases from a very low value to the critical value at a constant capture area ratio. Realistically, the flight inlet will operate with a supercritical margin, shown in Fig. 5.3, and never achieve the critical value, because the exit nozzle throat is oversized, as explained earlier. As the terminal normal shock moves upstream from the cowl lip, the total pressure recovery remains at about the critical level, the static pressure upstream from the cowl lip increases, and the capture area ratio

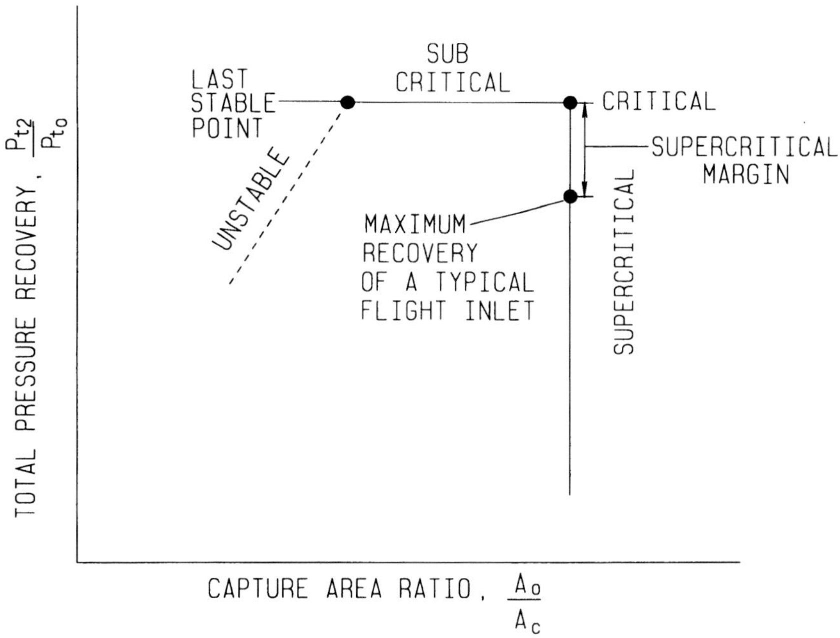

Fig. 5.3 Qualitative performance of typical inlet as a function of terminal normal shock position.

is reduced. If, at some subcritical position of the terminal normal shock, it becomes aerodynamically impossible for the inlet entrance static pressure to increase with increasing combustor static pressure, the terminal normal shock will become unstable, and the mean value of the total pressure recovery will decrease as shown by the dashed line in Fig. 5.3. The subcritical shock position at which the total pressure recovery curve slope changes is called the last stable point. At any given terminal normal shock position, the capture area ratio and total pressure recovery of the typical inlet vary with freestream Mach number, as shown in Fig. 5.4. Usually, only the critical values are plotted in this fashion, although the trend is the same for other shock positions. At all Mach numbers above the design value, the capture area ratio remains constant, but the total pressure recovery decreases with increasing Mach number. This trend is the consequence of two streams of air with different compression histories sharing the same passage. The slip planes, or vortex sheet, which separates the two streams, produce friction and mixing losses, and these losses increase with increasing Mach number as the wedge shock and compression fan move farther into the inlet so that more uncompressed air is captured. At all Mach numbers below the design value, the inlet spills air, and the spillage increases with decreasing Mach number as the wedge shock and compression fan move farther in front of the inlet. The corresponding total pressure recovery increases because both the wedge shock and normal shock recoveries increase with decreasing Mach number.

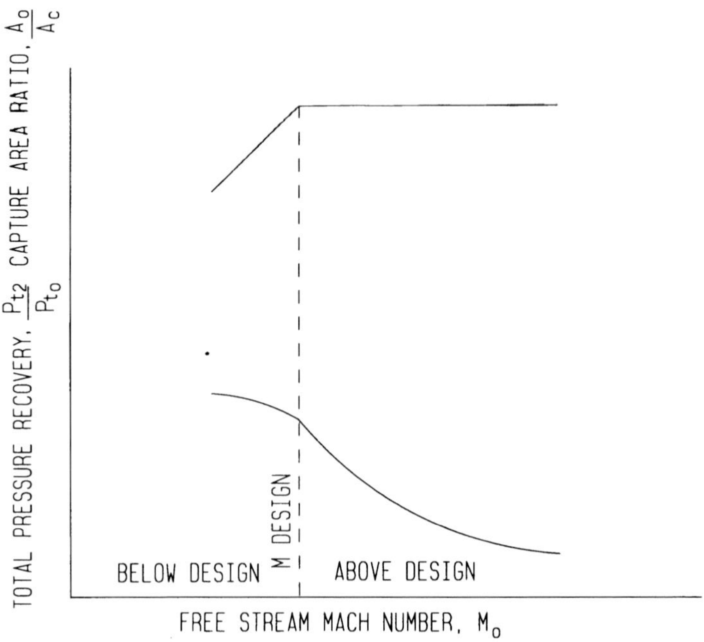

Fig. 5.4 Qualitative performance of typical inlet as a function of freestream Mach number.

5.3 Angle-of-Attack Effects

The discussion has thus far been limited to inlet operation at 0-deg angle of attack; i.e., the direction of the freestream velocity vector is coincident with the x axis of the inlet. The effect of a change in freestream flow direction relative to the inlet axis is shown in Fig. 5.5 for a relatively simple conical centerbody, flow-aligned cowl axisymmetric inlet. Figure 5.5a illustrates the symmetry of the flow picture during on-design operation at 0-deg angle of attack, and Fig. 5.5b shows how that symmetry is destroyed by rotating the inlet to angle of attack with the freestream flow direction.

It is a characteristic of conical flow at supersonic velocities that the shock wave rotates less than the surface to which it is attached. The amount of cone shock rotation relative to cone surface rotation for any cone half-angle at any supersonic freestream Mach number can be determined from Ref. 6, wherein the word "yaw" is used to describe the angle of rotation. Inasmuch as the flow picture is the same in either the yaw or pitch plane, the angle of yaw and the angle of attack are interchangeable descriptions of angle of rotation. Whether the cone is yawed or pitched, it would be expected that crossflow increases the surface Mach number on the leeward side compared with the unrotated value and decreases surface Mach number on the windward side. Concurrently, the leeward surface static pressure decreases and the windward surface static pressure increases compared with the unrotated value. Flow conditions on the leeward and windward

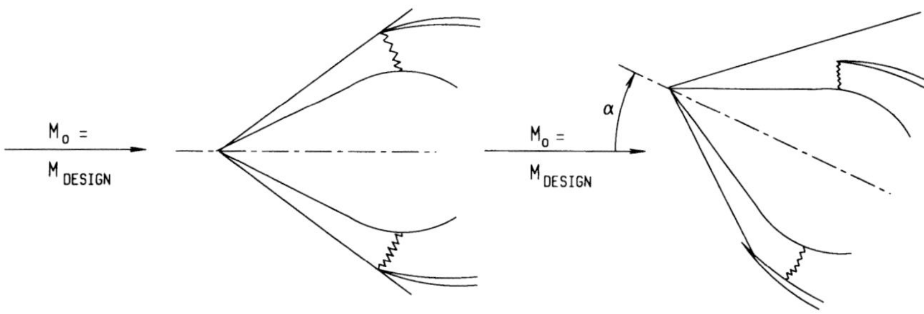

Fig. 5.5 Sketches of two operating conditions of an axisymmetric inlet with single-surface conical centerbody: a) 0 deg α; b) high value of α.

surfaces of a rotated cone can be determined by the method presented in Ref. 6. Results obtained with the analytical method have been proven to be very accurate compared with measured surface static pressures collected in wind-tunnel tests with cones at precisely defined angles of rotation. An alternate approach, somewhat simpler in application but somewhat less accurate than the method of Ref. 6, requires knowledge of the cone shock to cone surface yaw angle ratio and access to the conical flowcharts of Ref. 5. Development of the steps that comprise the alternate approach and a specific example of the use of the method are described later in the chapter.

Using the values and terminology of Ref. 6, the ratio of shock to cone yaw angle is plotted in Fig. 5.6 over a range of freestream Mach numbers from 2.0 to 4.0 for cone half-angles of 25 and 15 deg, which are the recommended choices for a single-surface conical inlet and a semi-isentropic surface initial cone, respectively. A 25-deg half-angle cone operating at Mach 3.0 has a shock to cone yaw angle of approximately 0.84. If the cone were rotated to a 10-deg angle of attack with the local flow direction, the shock would rotate to 8.4 deg, which is the effective angle of attack. The cone flowcharts of Ref. 5 can be used to find both leeward and windward surface flow conditions if the cone half-angle in the leeward wide is reduced by the effective angle of attack and the cone half-angle on the windward side is increased by the effective angle of attack. In the example case the 16.6-deg half-angle leeward side cone has a surface Mach number M_S of 2.43 and a surface static pressure ratioed to freestream total pressure, P_S/Pt_0, of 0.062. The 33.4-deg half-angle windward side cone has a surface Mach number of 1.66 and a static to total pressure ratio of 0.147. These compare with unrotated 25-deg half-angle cone surface values of $M_S = 2.05$ and $P_S/Pt_0 = 0.100$. The pressure differential between windward and leeward surfaces, i.e., $(P_S/Pt_0)_{\text{wind}} - (P_S/Pt_0)_{\text{lee}} = \Delta P_S/Pt_0$, is 0.085, which compares with a value of 0.088 using the method of Ref. 6. This difference is about 3.5%, which is within the normal range of accuracy of experimental measurement.

In light of the foregoing discussion of rotated cones, the flow picture of a conical inlet operating at a high angle of attack, as shown in Fig. 5.5b appears less intractable to an orderly interpretation. Because the conical

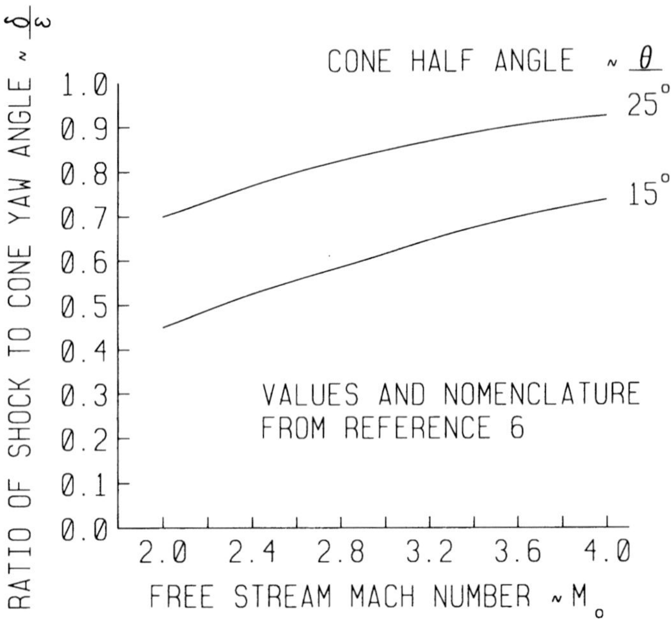

Fig. 5.6 Variation of shock yaw angle to cone yaw angle ratio with freestream Mach number for two cone half-angles.

shock wave "sees" a lower angle of attack than the inlet surfaces, the windward side of the shock sheet crosses the cowl lip plane in a manner akin to normal inlet operation above the design Mach number, whereas the lee side of the shock sheet lies forward of the cowl lip similar to inlet operation below design Mach number. Thus, the inlet must spill air on the lee side, and the amount of spillage must increase with increasing angle of attack. Because of crossflow, however, the lee side Mach numbers at the cowl lip plane are higher than design, and the windward side Mach numbers are lower than design. Based on the proposition that the flow will adjust to produce a constant strength shock, the windward side flow must accelerate to an average Mach number on the same order as the lee side value at the cowl lip plane before the terminal normal shock occurs. Thus, the terminal shock must be skewed so that critical total pressure recovery is achieved with the lee side of the shock in the throat, but with the windward side of the shock in the subsonic diffuser, akin to supercritical operation. With increasing angle of attack, the terminal normal shock must become more skewed, and the critical total pressure recovery of the inlet must continue to decrease. The rationale just described also applies to an axisymmetric inlet with a semi-isentropic spike, and the measured performance of such an inlet is shown in Fig. 5.7, where capture area ratio and critical total pressure recovery are plotted against inlet angle of attack. Data were collected in wind-tunnel tests of a scaled model operating at the design Mach number of the inlet. Tests at off-design Mach numbers

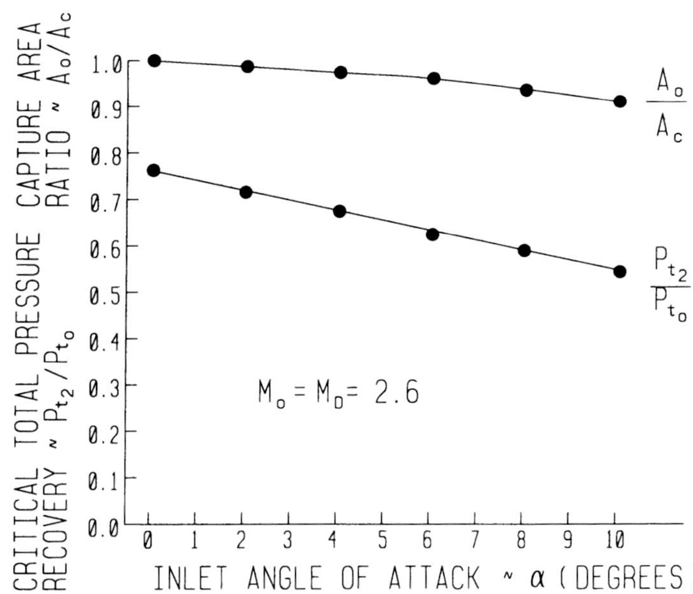

Fig. 5.7 Effect of angle of attack on the measured performance of an axisymmetric semi-isentropic inlet.

showed a similar pattern of decreasing performance with increasing angle of attack. Generally the loss of engine thrust associated with the decreased inlet performance precludes the selection of an axisymmetric inlet for any mission where maneuvering flight is required.

However, if an axisymmetric inlet can be shielded from the adverse crossflow, it is possible to realize an increase in performance with increasing angle of attack, as shown in Fig. 5.8. The shielding device, which is shown in the sketch, is a simple flat plate that produces a two-dimensional flowfield. Conditions upstream from the plate shock are termed infinity (∞), and local conditions at the inlet face are termed zero, which is consistent with the nomenclature already defined in previous chapters. Inspection of the flow picture reveals that the area A_∞ of the freestream tube captured by the inlet is larger than the area A_0 of the local stream tube, and that the ratio A_∞/A_0 increases with increasing angle of attack of the plate. Also, for any given value of M_∞ the value of M_0 will decrease with increasing angle of attack of the plate. As shown in Fig. 5.4, the value of the inlet critical total pressure recovery will increase with decreasing M_0. Although the total pressure recovery across the plate shock will decrease with increasing angle of attack at any M_∞ and with increasing M_∞ at any angle of attack above 0 deg, the product of the plate total pressure recovery and the overall inlet total pressure recovery at reduced M_0 will almost always show a net gain compared with the 0-deg angle-of-attack performance. Similarly, the solution to the continuity equation written between stations ∞ and 0 will show an increase in the cross-sectional area of the freestream tube captured

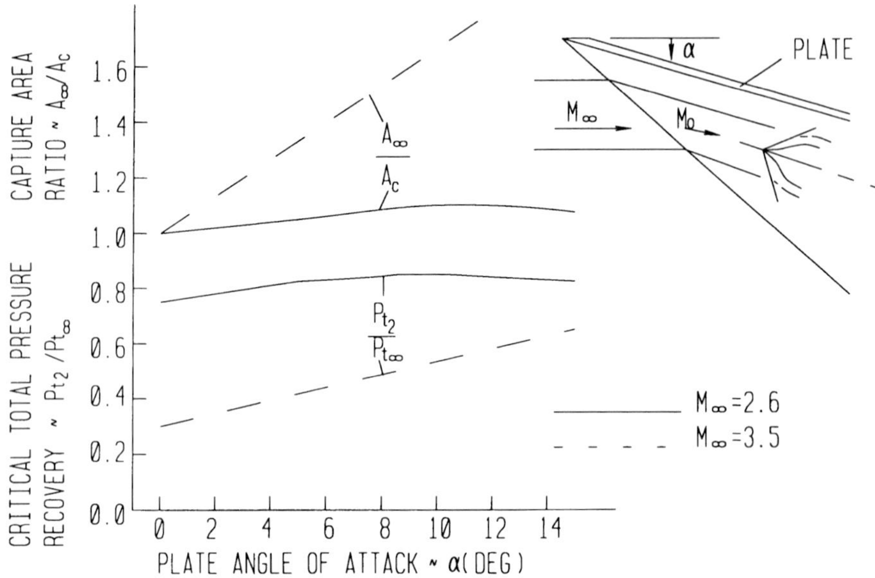

Fig. 5.8 Effect of a two-dimensional pressure field on the performance of an axisymmetric inlet.

by the inlet compared with the 0-deg condition. The effect of the two-dimensional field generated by a flat plate on the performance of the axisymmetric inlet (whose unprotected performance is given in Fig. 5.7) is plotted in Fig. 5.8 for a relatively low value of freestream Mach number, $M_\infty = 2.6$, and a relatively high value, $M_\infty = 3.5$. At 0-deg plate angle of attack at the low Mach number, $M_\infty = M_0 = 2.6$, the inlet is operating on design, and the capture area and critical total pressure recovery values are the same as those plotted in Fig. 5.7. As plate angle of attack increases, both capture and recovery increase, reach a maximum in the vicinity of 10 deg, and begin to decrease between 10 and 15 deg. At $M_0 = 3.5$, both capture and recovery exhibit a constant increase in value between 0- and 15-deg plate angle of attack. The modest gains in angle-of-attack performance at low Mach numbers, usually near the design value, and the significant gains at high Mach numbers, usually near the upper end of the operating range, are typical of properly protected or oriented fixed-geometry inlets of any family. An examination of the qualitative performance curves in Fig. 5.4 shows the basic cause of these performance differences. As local Mach number M_0 decreases below the design value, the inlet begins to spill air so that the increase in A_∞/A_0 is offset by decreases in A_0/A_C. Also, the increase in Pt_2/Pt_0 with decreasing M_0 is offset by the increase in pressure loss across the plate shock. Conversely, at the high freestream Mach numbers, decreases in the local Mach number M_0 with increasing α produce substantial increases in Pt_2/Pt_0, which outweigh the increased shock losses. Moreover, the local capture area A_0/A_C remains at unity over the entire local Mach number range, so that the gains in A_∞/A_0 are not offset by any spillage.

In the practical application of an axisymmetric inlet to a missile, the protective device would most likely be a wing or even the vehicle body. However, a properly oriented two-dimensional inlet could provide its own favorable pressure field. Properly oriented means that the inlet is compressing in a downward direction in the pitch plane so that positive-angle-of-attack operation increases the external compression compared with the 0-deg value. A downward compression semi-isentropic two-dimensional inlet is shown in the sketches in Fig. 5.9. The external flow pictures for the basic inlet at the design Mach number $M_0 = M_D = 2.5$ at 0 deg and at 10-deg angle of attack is given in Figs. 5.9a and 5.9b, respectively, and for the basic inlet plus a shroud at 10-deg angle of attack on Fig. 5.9c. The shroud consists of a flat plate with triangular side walls swept back to form a contiguous boundary with the basic inlet side-wall leading edge. Inspection of the three sketches in Fig. 5.9 reveals that the height and therefore the area A_0 of the freestream tube captured by the inlet increases with increasing angle of attack and with increasing shroud length for any angle of attack above 0 deg. Closer inspection of Fig. 5.9b reveals that the wedge shock is curved because of a multitude of intersections with the isentropic compression waves. Because of the shock curvature, the basic inlet spills air at Mach numbers above the value where the two-dimensional wedge shock impinges on the cowl lip. The addition of a shroud to the basic inlet eliminates the curved shock problem. Shroud length is a variable, but a practical maximum would probably be set by plate shock impingement on cowl lip at some high angle of attack (e.g., 15 deg) at some high operating Mach number (e.g., 4.0).

The angle-of-attack performance of the basic two-dimensional downward-compressing inlet and the effect of the shroud on angle-of-attack performance are given in Fig. 5.10. The performance is presented at the same two Mach numbers (i.e., 2.6 and 3.5) used previously in Fig. 5.8 for the axisymmetric inlet protected by the flat plate in order to compare trends. A comparison of magnitudes is inappropriate because the inlets were designed for different compression levels at different design Mach

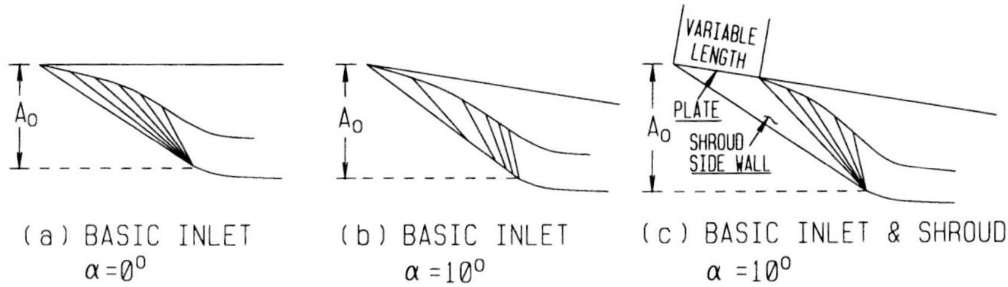

Fig. 5.9 Sketches of three operating conditions of a two-dimensional inlet compressing in downward direction in the pitch plane: a) basic inlet (α = 0 deg); b) basic inlet (α = 10 deg); c) basic inlet and shroud (α = 10 deg).

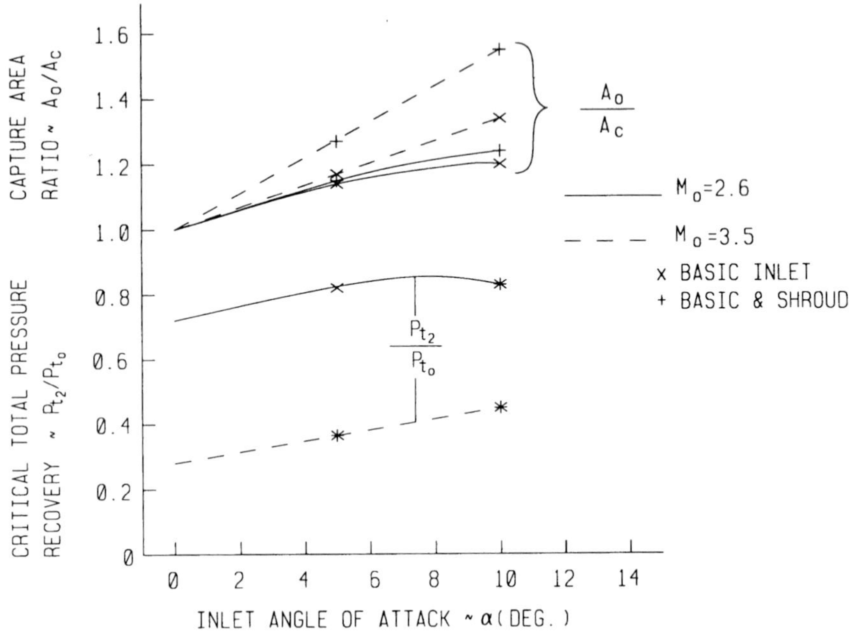

Fig. 5.10 Estimated performance of a downward-compressing two-dimensional inlet with and without a precompression shroud.

numbers. The two-dimensional inlet performance at any angle of attack at either Mach number is better than the 0-deg level, but the performance gains are much more significant at the higher Mach number than at the lower Mach number. These trends are identical to those shown and are explained previously for the protected axisymmetric inlet. The incorporation of a shroud with the basic inlet has little, if any, impact on the critical total pressure recovery, at either Mach number, but does have a measurable effect on capture area ratio at the lower Mach number and a substantial effect at the higher Mach number. The bulk of any improved capture area ratio with the shroud is a result of simple geometry, as shown in Fig. 5.9, and the longer the shroud, the greater the capture area. However, some of the improvement is traceable to the restoration of two-dimensional flow upstream from the basic inlet and the elimination of curved shocks. In effect, the shroud serves the same purpose for the two-dimensional inlet as the flat plate (wing) does for the axisymmetric inlet. Usually, the shroud is an integral part of the two-dimensional inlet where the wing is a separate entity, but a flat plate could be an integral part of an axisymmetric inlet, and a basic two-dimensional inlet could be protected by a wing or a vehicle body. The primary difference is that an unprotected basic two-dimensional inlet can achieve improved performance with increasing angle of attack, whereas the unprotected axisymmetric inlet will always show decreasing performance. Whether protected by shrouds or wings or not, all inlets will suffer performance losses during operation at negative angles of attack, and the trends will be similar to those shown in Fig. 5.7 for the unprotected axisymmetric inlet.

One facet of the flow picture not obvious in Fig. 5.9 for the downward-compressing two-dimensional inlet at 10-deg angle of attack is that the supersonic flow has been turned beyond the limiting value given in Fig. 2.13. The consequences of excessive compressive turning are not apparent until the wedge shock, in the case of the basic inlet, or the shroud shock crosses the cowl lip plane and uncompressed air enters the inlet. The resulting flow picture is shown in Fig. 2.13, and the concomitant performance loss is highlighted in Fig. 5.11 for a shrouded two-dimensional inlet operating at 10-deg angle of attack. Below Mach 3.0, where the shroud shock crosses the cowl lip plane, the measured capture area ratio matches the estimated values quite closely. Above Mach 3.0, the measured capture area ratio is considerably lower than estimated, and engine performance is penalized in two ways: 1) The air mass flow available for acceleration through the nozzle is reduced; and 2) the air spilled around the cowl increases the external drag. The angle of attack at which the compressive turning limit is initially exceeded can be determined for any two-dimensional inlet by means of Fig. 2.13 and the discussion contained in Chapter 2. The consequences of excessive turning can be postponed by increasing the shroud length so that the shroud shock at any angle of attack crosses the cowl lip at a higher Mach number.

5.4 Control Techniques

In Chapter 1 the flow-through ramjet engine with in-line components (inlet, combustor, and nozzle) was used as the example in the discussion of the Brayton cycle and in the derivation of the thrust equation. The flow-through ramjet was chosen for two reasons: 1) It has an inherent simplicity that lends itself to a lucid explanation of form and function; and 2) it has been proven in operational use as the sustain engine on the BOMARC and TALOS supersonic missiles. Both missile engines were liquid-fueled, and thrust was varied by regulating fuel flow rate. As shown in the development of the engine performance equations presented in Chapter 1 and in the discussion of inlet operational characteristics contained in this chapter, increasing the fuel flow rate toward the stoichiometric value increases the back pressure in the combustor, which moves the terminal normal shock toward the critical location in the inlet. The normal shock movement increases the total pressure recovery across the engine, which increases engine thrust.

Fuel flow from a containment tank to the engine injectors is regulated or metered by means of a throttle valve, which can be opened or closed by means of an actuation device. Actuation devices can consist of bellows, diaphragms, pistons, servovalves, electronic (torque motors, stepper motors) or fluidic, or some combination of the these devices. The actuation device causes the throttle valve to move in response to a signal that can be generated onboard the missile or telemetered from some source outside the

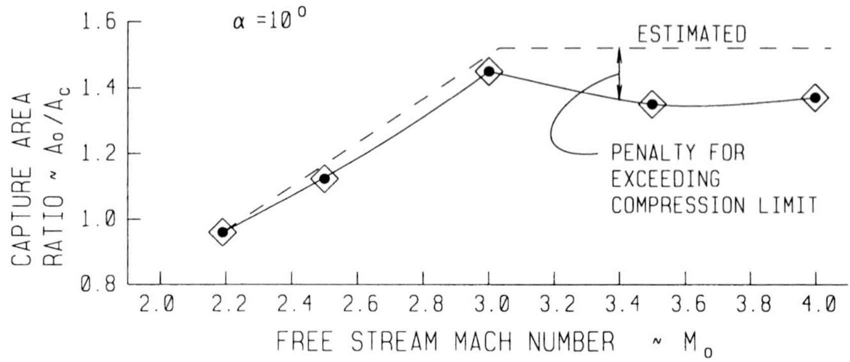

Fig. 5.11 Effect of excessive compressive turning on measured performance of a downward-compressing two-dimensional inlet with precompression shroud.

missile. The amount and direction of throttle valve movement in response to the signal is determined by the flow logic designed into the fuel controller. One simple form of flow logic is the altitude scheduled fuel controller, which uses an actuation diaphragm. One side of the diaphragm is evacuated, and the other side is connected to a static pressure tap located on the missile body surface. As the missile climbs, the diaphragm deflects to close the valve and reduce the fuel flow rate commensurate with the decrease in air mass flow swallowed by the inlet. The fuel controller can be programmed to respond to a variety of signals, including altitude, Mach number, angle of attack, combustion chamber temperature, or even time.

In order to preclude inlet operation in the subcritical regime where engine net thrust is reduced below the peak value, the onset of buzz can cause structural damage, and engine blowout is possible, the flow controller can be programmed to override any other instruction and reduce fuel flow when the terminal normal shock approaches the inlet throat. A static tap strategically located on a wall of the subsonic diffuser near the throat can be used to send the signal to reduce fuel flow. The static tap senses the pressure rise associated with the advancing shock wave and begins the process of stopping the shock at some supercritical position, usually at 90–95% of the critical total pressure recovery value. The selection of the best static tap for the shock positioner loop of the fuel controller is usually determined during inlet model tests in a wind-tunnel. Other more sophisticated techniques for shock positioner control can be used, but each one requires a map of inlet performance such as those depicted in Fig. 5.2, plus performance changes caused by angle-of-attack and yaw operation.

6 PERFORMANCE ESTIMATION

6.1 Total Pressure Recovery

The on-design operating condition, depicted in Fig. 5.1 for a typical ramjet inlet, provides the foundation for all performance-estimating procedures, including capture area ratio and drag, which will be discussed in sequence after the total pressure recovery parameter. As described in previous chapters, at the design condition all flow properties are known at the key inlet stations identified in Fig. 6.1. Overall total pressure recovery is the product of the recoveries across the three constituents of an inlet (i.e., supersonic diffuser, throat section, and subsonic diffuser). Mathematically,

$$\frac{Pt_2}{Pt_0} = \frac{Pt_1}{Pt_0}\frac{Pt_{1'}}{Pt_1}\frac{Pt_2}{Pt_{1'}} \tag{6.1}$$

Comparisons of estimated and measured values of inlet total pressure recovery have shown that consistently high estimating accuracy can be achieved if the following procedures are used in the solution of Eq. (6.1):

1) Flow through the supersonic diffuser is treated as inviscid, and the total pressure recovery Pt_1/Pt_0 is synonymous with the recovery across the leading-edge wedge, or cone, shock at the design freestream Mach number. Oblique shock recoveries for wedges and cones are plotted in Ref. 6.

2) Flow through the throat region is also treated as inviscid, and the total pressure recovery $Pt_{1'}/Pt_1$ is the recovery across a normal shock at the Mach number M_1. Normal shock recoveries are listed in Refs. 1 and 2.

3) Flow through the subsonic diffuser is treated as viscous, and the total pressure recovery $Pt_2/Pt_{1'}$ is represented by the empirically determined value of 0.90, which incorporates all of the boundary-layer-related losses across the entire inlet.

The procedures described earlier have been applied to the three example inlets designed in Chapter 3, and the results are given in Table 6.1.

The estimating procedures defined and illustrated earlier apply only to inlets, such as the example cases, which have been designed in accordance with the precepts promulgated in Chapters 2 and 3 to achieve peak overall

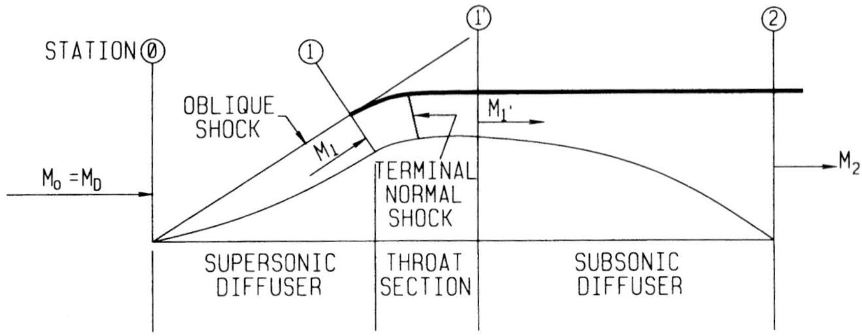

Fig. 6.1 Sketch of on-design inlet showing key stations for performance-estimating purposes.

total pressure recovery Pt_2/Pt_0. If an inlet is designed with a throat length shorter than the shock train length, the peak recovery would be reduced by an amount consistent with the results of Ref. 14 as shown in part in Fig. 3.4. An additional reduction in peak recovery would be observed if the equivalent conical half-angle of the subsonic diffuser exceeds a value on the order of 8 deg. The amount of the reduction would best be determined in experimental tests of the complete inlet configuration. Total pressure losses caused by dumping the flow from a smaller to a larger area will be calculated in the next chapter in the description of inlet installations on vehicles.

During inlet operation below design Mach number with the normal shock in the critical position, Eq. (6.1) is still valid, but the supersonic diffuser and throat section recovery values change. As shown in Fig. 6.2a, the oblique shock generated by the leading-edge wedge or cone moves forward from the cowl lip, and the forward movement indicates a decrease in shock strength. There are two consequences of a reduction in oblique shock strength: The total pressure recovery across the shock Pt_1/Pt_0 increases; and the Mach number in the wedge or cone field decreases. If the flow is compressed an additional amount by isentropic turning through a fixed number of degrees, the final supersonic Mach number M_1 at the beginning of the throat section will be less than the design value, and the strength of the normal shock will be decreased. Consequently, the total pressure recovery across the normal shock $Pt_{1'}/Pt_1$ increases as the freestream Mach number decreases below the design value. Experimental results seem to confirm a constant value (0.90) of the empirically derived total pressure recovery $Pt_2/Pt_{1'}$ across the subsonic diffuser. The method for solving Eq. (6.1) to determine inlet performance below design Mach number is illustrated in Table 6.2 using the Mach 3.0 design inlet from Chapter 3 as an example. The on-design Mach number results from Table 6.1 are repeated for comparison.

If an axisymmetric inlet is the subject of analysis, it is necessary to determine the average Mach number in the initial cone field before turning

Table 6.1 Calculation of on-design total pressure recovery for three example inlets

$M_0 = M_D$	M_1	5-deg Wedge at M_0 Pt_1/Pt_0	×	Normal shock at M_1 $Pt_{1'}/Pt_1$	×	Empirical value $Pt_2/Pt_{1'}$	=	Pt_2/Pt_0
2.0	1.47	1.000		0.939		0.90		0.845
3.0	1.99	0.994		0.726		│		0.650
4.0	2.52	0.990		0.491		↓		0.438
Values taken from Fig. 3.8		Values taken from Ref. 6		Values taken from Ref. 1				Values plotted in Fig. 3.9

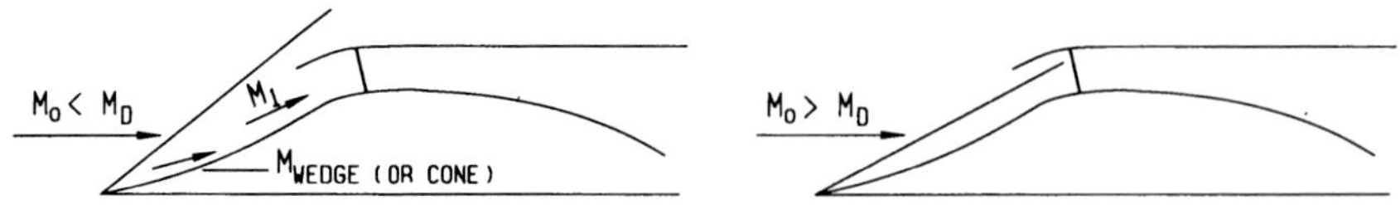

Fig. 6.2 Sketches of inlet operating at critical recovery off-design Mach number with key parameter identified: a) below design; b) above design.

the flow through the Prandtl-Meyer process. Reference 6 gives plots of cone surface Mach number as a function of freestream Mach number and cone deflection angle. Reference 6 also shows conical shock-wave angle for the same variables. The Mach number behind the cone shock is always higher than the cone surface value and can be readily determined from the fact that there is a wedge deflection angle that produces the same oblique shock wave angle as the half-cone. For example, a 15-deg half-angle cone at Mach 3.0 has a shock-wave angle of 25.3 deg (from Ref. 6), which is the same as a 7.6-deg wedge at the same freestream Mach number. Thus, the Mach number behind the conical shock is the same as that in a 7.6-deg wedge flowfield—roughly 2.62 from the appropriate plot in Ref. 6. The cone surface Mach number from Ref. 6 is 2.50, and the average Mach number in the cone field is $2.62 + 2.50 = 5.12/2 = 2.56$.

A more direct approach to the calculation of total pressure recovery below design Mach number is facilitated by the tangent wedge (cone) approximation. The relationship between the supersonic Mach number M_1 at the beginning of the throat section and the freestream Mach number M_0 at the design condition can be achieved by a single two-dimensional (conical) deflection angle that can be determined from the appropriate plots in Ref. 6. That particular deflection angle is used at freestream Mach numbers below the design value to determine corresponding values of M_1. For example, the Mach 3.0 design inlet from Table 6.2 has an on-design $M_1 = 1.99$. From the Ref. 6 plot of Mach numbers before and after a plane shock, a deflection angle slightly less than 20 deg would produce the proper relationship. At $M_0 = 2.5$, the deflection angle yields an $M_1 \cong 1.645$, and at $M_0 = 2.0$, $M_1 \cong 1.205$. The latter values of M_1 are somewhat higher than the values determined from the initial wedge plus isentropic turning process shown in Table 6.2, but the total pressure recovery differences between the two methods are well within the experimental accuracy achieved in wind-tunnel tests.

During critical inlet operation above design Mach number, Eq. (6.1) is invalid. Fig. 6.2b shows that the oblique shock generated by the leading-edge wedge or cone has moved inside the cowl lip so that uncompressed freestream air is entering the inlet. Downstream from the cowl lip plane the uncompressed flow must share the enclosed passage with the flow that has been compressed by the wedge (cone) and isentropic turning process. The slip plane, or vortex sheet, which separates the two streams, produces friction and mixing losses that have been analytically intractable. Based on results compiled from many inlet tests, the critical total pressure recovery

Table 6.2 Calculation of critical total pressure recovery below design Mach number for a Mach 3.0 design example inlet

	5-deg Deflection			Prandtl-Meyer turning through 18.6 deg				5-deg Wedge at M_0		Normal shock at M_1		Empirical value		
	M_0	M_{wedge}	→	ν_w	ν_1	→	M_1	$\frac{Pt_1}{Pt_0}$	×	$\frac{Pt_{1'}}{Pt_1}$	×	$\frac{Pt_2}{Pt_{1'}}$	=	$\frac{Pt_2}{Pt_0}$
Design	3.0	2.75		44.69	26.09		1.99	0.994		0.726		0.90		0.650
	2.5	2.29		34.03	15.43		1.62	0.996		0.888		↓		0.796
	2.0	1.82		21.30	2.70		1.16	0.998		0.996		↓		0.895
		Ref. 6		Ref. 1 or 6			1 or 6	6		1				

of any inlet operating above design freestream Mach number may be determined with reasonable accuracy from the following empirical expression:

$$\left(\frac{Pt_2}{Pt_0}\right)_{M_0} = \left(\frac{Pt_2}{Pt_0}\right)_{M_D} e^{b(M_D - M_0)} \tag{6.2}$$

where $(Pt_2/Pt_0)_{M_0}$ is the critical total pressure recovery at any freestream Mach number above the design value; $(Pt_2/Pt_0)_{M_D}$ the critical total pressure recovery at the design Mach number M_D; e = 2.718; b = 1.018 for isentropic surfaces with zero throat length, = 0.685 for particular isentropic inlet with 24 channel heights of constant area throat, = 0.855 for single cone inlets (zero throat length).

Obviously, the slope of the critical total pressure recovery curve as a function of freestream Mach number above the design value for an isentropic inlet can be controlled by throat length: The longer the throat, the more shallow the slope. (It is speculated that similar results can be achieved by proper application of bleed in lieu of throat length.) The slope for the zero-throat-length isentropic inlet has been confirmed in many tests of both two-dimensional and axisymmetric isentropic inlets designed at a number of different Mach numbers and operated over a wide range of freestream Mach numbers above the design values. In the absence of sufficient data to define the exact shape of the above design total pressure recovery curve as a function of the throat length, a tentative approximation is shown in Fig. 6.3 where a straight line connects the well-documented zero-throat-length point and the throat length of 24 channel heights with a slope value determined by a single test. The curve for conical inlets consists of a single

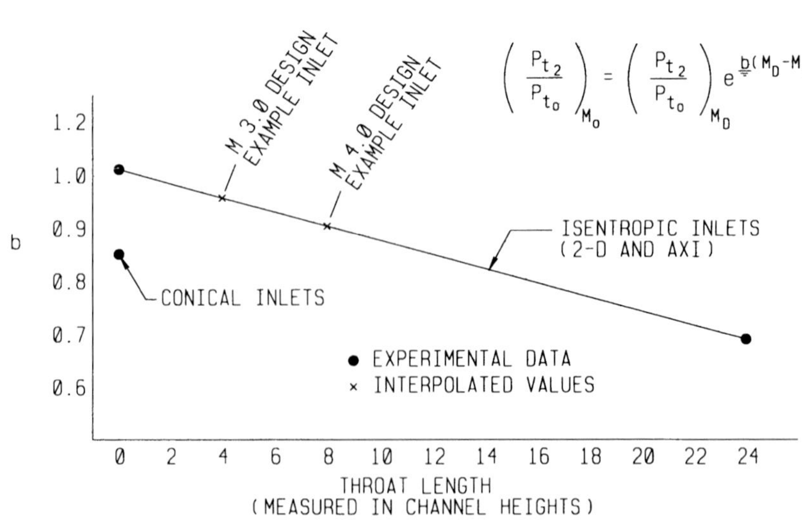

Fig. 6.3 Tentative relationship between the slope of the critical total pressure recovery curve above inlet design Mach number and inlet throat length.

Table 6.3 Calculation of critical total pressure recovery above design Mach number for three example inlets

Inlet design Mach number M_D		2.0	3.0	4.0
Throat length (Fig. 3.8)		0	4	8
Slope exponent, b (Fig. 6.3)		1.018	0.96	0.90
Total pressure recovery at M_D (Table 6.1)		0.845	0.650	0.438
Freestream Mach number	M_0	Critical total pressure recovery at M_0		
	2.0	0.845	—	—
	2.5	0.508	—	—
	3.0	0.305	0.650	—
	3.5	0.184	0.402	—
	4.0	—	0.249	0.438
	4.5	—	0.154	0.279
	5.0	—	—	0.178

point identified in Fig. 6.3 at zero throat length. No data have been published at other throat lengths.

Equation (6.2) has been applied to the three example inlets designed in Chapter 3, and the results are presented in the Table 6.3.

To facilitate comparisons of the slopes of the curves defined by the values presented in Table 6.3, the critical total pressure recoveries of the three inlets are plotted vs freestream Mach number in Fig. 6.4 on semilog paper where the exponential equation yields straight lines.

For operating conditions other than critical, no formal estimating procedures are presented in the literature on ramjet inlets. Subcritical operation, as depicted in Fig. 6.5a, is almost always circumvented by design techniques, operational restrictions, or a shock position control. If the inlet should enter the subcritical regime where the flow is stable, the Mach number M_1 will be about the same as the critical value; thus the total pressure recovery will be practically unchanged, as shown in Fig. 5.3. If the flow becomes unstable, the solution to the flow equations is of no practical value to the inlet designer. During operation at supercritical conditions, as illustrated in Fig. 6.5b, the flow accelerates through the throat region to a supersonic Mach number M_1 in the subsonic diffuser, where a normal shock that is stronger than the critical value occurs. As the normal shock moves farther downstream in the subsonic diffuser, the shock strength increases, and the total pressure recovery of the inlet decreases, as shown in Fig. 5.3. The flow area in the subsonic diffuser at the normal shock position A_1 ratioed to the throat flow area A_T can be used in Eq. (1.35) to solve for the value of M_1, and the corresponding normal shock total pressure recovery is given in Refs. 1 and 2. The position of the normal

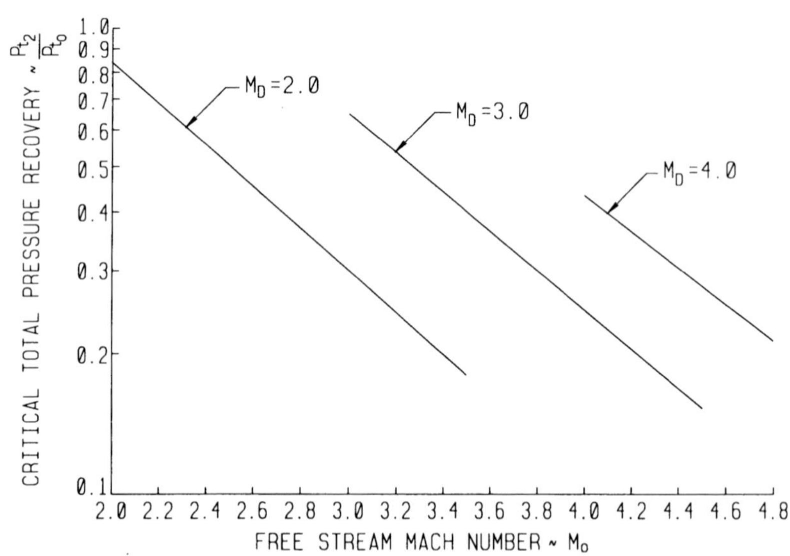

Fig. 6.4 Estimated critical total pressure recovery for three two-dimensional isentropic inlets operating above design Mach number.

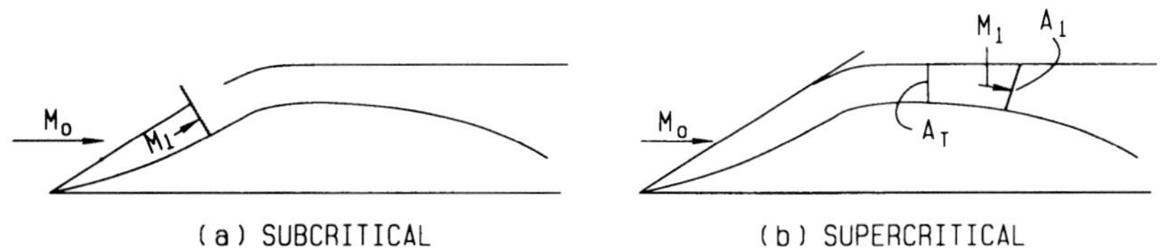

Fig. 6.5 Sketches of inlet operating with the terminal normal shock upstream and downstream from the throat section: a) subcritical; b) supercritical.

shock in the subsonic diffuser is determined by the shock strength required to satisfy the continuity [Eq. (1.26)] in conjunction with the companion combustor total pressure recovery, Pt_4/Pt_2, which was defined and discussed in Chapter 1.

Ramjet-propelled air-launched missiles with high maneuverability requirements may feature an inlet protected by a shroud, wing, or vehicle body, and one popular inlet choice is a downward-compressing two-dimensional configuration with an integral shroud, as sketched in Fig. 5.9c. As described in Chapter 5, the incorporation of a shroud with a basic two-dimensional inlet has little impact on the critical total pressure recovery vs angle-of-attack curves at two selected freestream Mach numbers, but can cause a substantial enhancement of capture area characteristics at the higher Mach number, where the angle-of-attack engine thrust demands are usually greatest. The shrouded downward-compressing two-dimensional inlet has been adopted as the model for explaining angle-of-attack performance-estimating procedures, and the key stations required in the process are illustrated in the sketch in Fig. 6.6.

During angle-of-attack operation, the inlet critical total pressure recovery is the product of the individual recoveries across four elements (i.e.,

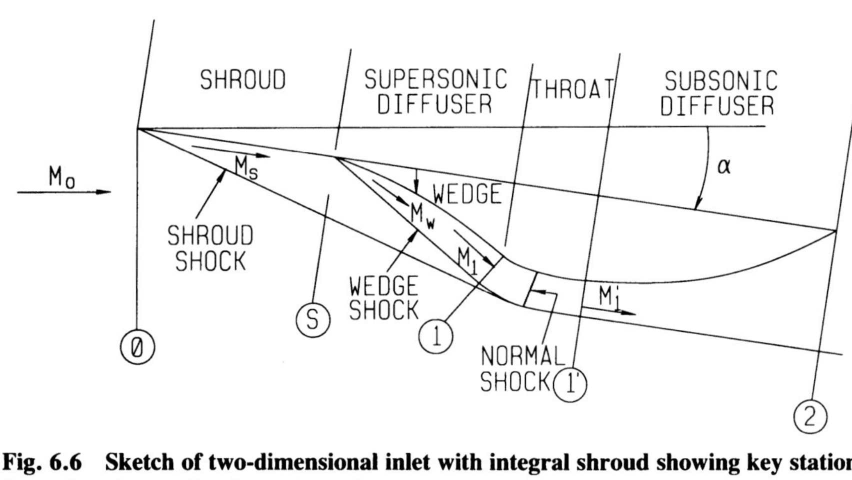

Fig. 6.6 Sketch of two-dimensional inlet with integral shroud showing key stations for estimating angle-of-attack performance.

shroud, supersonic diffuser, throat section, and subsonic diffuser):

$$\frac{Pt_2}{Pt_0} = \frac{Pt_s}{Pt_0}\frac{Pt_1}{Pt_s}\frac{Pt_{1'}}{Pt_1}\frac{Pt_2}{Pt_{1'}} \tag{6.3}$$

At 0-deg angle of attack ($\alpha = 0$ deg), the flat-plate element of the shroud produces no shock; thus, $Pt_S/Pt_0 = 1.00$ and Eq. (6.3) reverts to Eq. (6.1). At any positive angle of attack α, the shroud flat-plate deflects the freestream flow through an angle equal to the angle of attack. Both the shroud shock total pressure recovery Pt_S/Pt_0 and the shroud Mach number M_S can be found in the appropriate curves of Ref. 6. The other constituent total pressure recoveries and the other key Mach numbers can be found by the same process previously described for 0-deg angle-of-attack operation. Although Fig. 6.6 shows the shroud shock and the wedge shock (and, by implication, the compression fan focal point) coalescing at the cowl lip, this flow condition will occur only at one combination of freestream Mach number M_0 and angle of attack α for a given shroud length.

The procedures described earlier have been used to calculate the performance of a particular inlet at three operating angles of attack as shown in Fig. 6.7. The basic inlet is a downward-compressing two-dimensional configuration with a 5-deg leading-edge wedge followed by 18.5 deg of isentropic turning to a final compression surface angle of 23.5 deg and zero throat length. The wedge shock and compression fan focal point coalesce on the cowl lip at the freestream design Mach number $M_D = 2.5$ at 0-deg angle of attack. The integral shroud consists of a flat-plate that has a length equal to 63% of the basic inlet (supersonic diffuser) length and side walls cut along the line connecting the leading edge of the flat-plate and the cowl lip. The included angle of the side plate is 17.61 deg, as shown in Fig. 6.7a. At 0-deg angle of attack, the shroud flat-plate generates a Mach line that has an angle μ with the horizontal. There are no changes in flow properties across a Mach line; hence, the shroud has no effect on basic inlet performance at 0-deg angle of attack. As the inlet is rotated to angles of attack of 5 and 10 deg (Figs. 6.7b and 6.7c, respectively) the shroud flat-plate generates shocks with wave angles designated by θ_S. The changes in flow properties across the shroud shock and the impact on the other key elements have been determined by Eq. (6.3), and the results are summarized in Table 6.4 over a range of freestream Mach numbers at each of the three angles of attack.

Certain aspects of Table 6.4 are worthy of further discussion. It should be noted that the inlet design condition (i.e., wedge shock and compression fan focal point on cowl lip) at 0-deg angle of attack is achieved with a specific combination of Mach numbers: M_S, M_W, and M_1. As the angle of attack is increased, the design condition Mach number combination shifts to higher freestream Mach numbers, so that, in general, inlet critical total pressure recovery increases with increasing angle of attack at any freestream Mach number.

A previous statement that the shroud has no effect on basic inlet performance at 0-deg angle of attack is borne out mathematically in Table 6.4

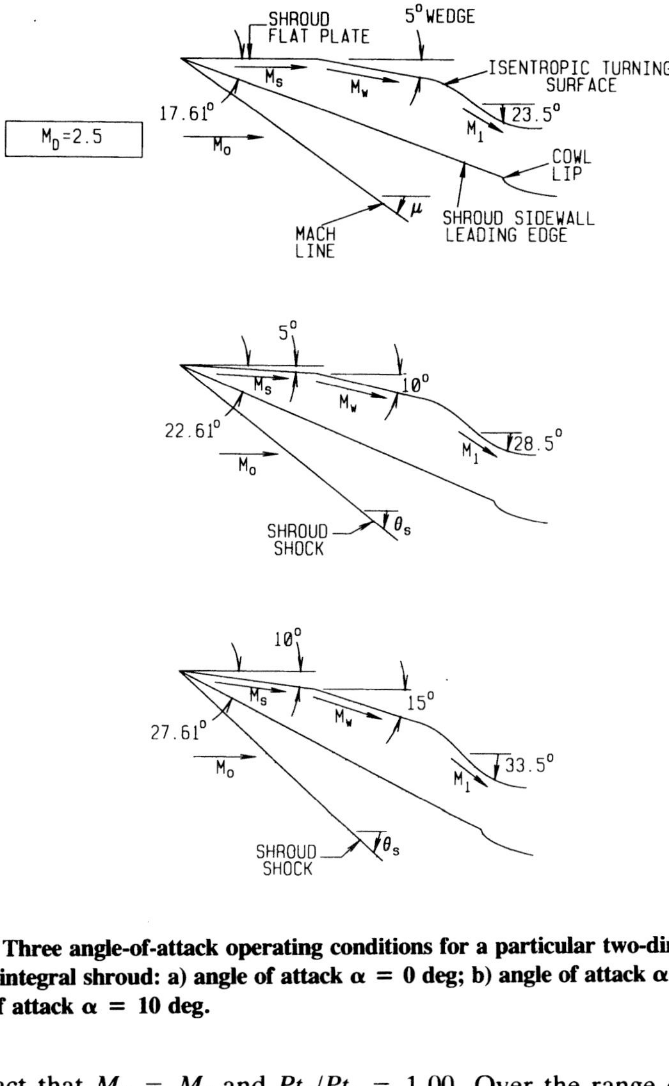

Fig. 6.7 Three angle-of-attack operating conditions for a particular two-dimensional inlet with integral shroud: a) angle of attack α = 0 deg; b) angle of attack α = 5 deg; c) angle of attack α = 10 deg.

by the fact that $M_S = M_0$ and $Pt_S/Pt_0 = 1.00$. Over the range of Mach numbers of 2.0–4.0 at 0-deg angle of attack, the Mach line generated by the shroud moves from a position well forward of the cowl lip to a position inside the cowl lip. The Mach line lies on the cowl lip at a freestream Mach number of 3.30, as noted on the top section of Table 6.4. It was expected that the position of the Mach line at the design condition would not influence the slope of the above design recovery curve, which is determined by the exponential equation shown in the table. However, at other angles of attack, where the shroud generated a shock, it was anticipated that the shock position relative to the cowl lip at the design condition could influence the angle and strength of the vortex sheet and hence the slope of the recovery curve. Limited experimental data obtained with a shrouded down-

Table 6.4 Calculation of critical total pressure recovery for a shrouded two-dimensional inlet at three angles of attack

$\alpha = 0$ deg

$M_D = 2.5$
Shroud Mach line $\mu = 23.58$ deg at design M
Shroud Mach line on cowl lip at $M_0 = 3.30$

	M_0	0-deg Deflect	M_S	5-deg Deflect	M_W	P-M turn 18.5 deg	M_1	0-deg Wedge at M_0: $\frac{Pt_S}{Pt_0}$	×	5-deg Wedge at M_S: $\frac{Pt_1}{Pt_S}$	×	NS at M_1: $\frac{Pt_{1'}}{Pt_1}$	×	Empirical: $\frac{Pt_2}{Pt_{1'}}$	=	$\frac{Pt_2}{Pt_0}$
Design—	2.5		2.5		2.29		1.624	1.000		0.995		0.886		0.90		0.793
Below design	2.2		2.2		2.01		1.375	↓		0.998		0.964		↓		0.866
	2.0		2.0		1.82		1.180	↓		0.999		0.995		↓		0.895
Above design	2.6 4.0		$\left(\frac{Pt_2}{Pt_0}\right)_{M_0} = 1.018\,(2.5\text{-}M_0)$ 0.793e													

$\alpha = 5$ deg

"Design" $M_0 = 2.72$
Shroud shock $\theta_S = 25.4$ deg at design M
Shroud shock line on cowl lip at $M_0 = 3.08$

	M_0	5-deg Deflect	M_S	5-deg Deflect	M_W	P-M turn 18.5 deg	M_1	5-deg Wedge at M_0: $\frac{Pt_S}{Pt_0}$	5-deg Wedge at M_S: $\frac{Pt_1}{Pt_S}$	NS at M_1: $\frac{Pt_{1'}}{Pt_1}$	Empirical: $\frac{Pt_2}{Pt_{1'}}$	$\frac{Pt_2}{Pt_0}$
Design—	2.72		2.50		2.29		1.624	0.995	0.995	0.886	0.90	0.789

	M_0	M_S	M_W	M_1	Pt_S/Pt_0	Pt_1/Pt_S	$Pt_{1'}/Pt_1$	$Pt_2/Pt_{1'}$	Pt_2/Pt_0
Below — Design	2.5	2.29	2.10	1.455	0.995	0.998	0.943	0.90	0.843
	2.2	2.01	1.83	1.181	0.999	1.000	0.994	0.90	0.894
	2.0	1.82	1.64	1.000	1.000	1.000	1.000	0.90	0.90
Above — design	2.8	$\left(\frac{Pt_2}{Pt_0}\right)_{M_0} = 1.018\,(2.72\text{-}M_0)$							
	4.0	$= 0.789e$							

α = 10 deg

"Design" M_0 = 3.0
Shroud shock θ_S = 27.5 deg at design M
Shroud shock on cowl lip at M_0 = 2.98

	M_0	10-deg Deflect M_S	5-deg Deflect M_W	P-M turn 18.5 deg M_1	10-deg Wedge at M_0 Pt_S/Pt_0	5-deg Wedge at M_S Pt_1/Pt_S	NS at M_1 $Pt_{1'}/Pt_1$	Empirical $Pt_2/Pt_{1'}$	Pt_2/Pt_0
Design—	3.0	2.50	2.29	1.624	0.962	0.994	0.886	0.90	0.763
Below — Design	2.5	2.085	1.90	1.26	0.975	0.998	0.986	↓	0.864
	2.2	1.82	1.64	1.00	0.982	1.000	1.000	↓	0.884
	2.0	1.635	1.46	1.00	0.985	1.000	1.000	↓	0.886
Above — design	3.2	$\left(\frac{Pt_2}{Pt_0}\right)_{M_0} = 1.018\,(3.0\text{-}M_0)$							
	4.0	$= 0.763e$							

Inlet lines sketched in Fig. 6.7.

ward-compressing inlet model operating above the design Mach number at several angles of attack show that the recovery curve slope is unchanged by shroud shock location. These data suggest that the length of the shroud will have no effect on the critical total pressure recovery characteristics of a basic two-dimensional downward-compressing inlet, except that increasing shroud length is synonymous with increasing viscous losses. Therefore, the exponential equation used for the above design calculation at 0-deg angle of attack in Table 6.4 applies to other angles of attack with appropriate changes in the design values to accommodate the shift in the design condition Mach number.

Results given in tabular form or implied from the equations in Table 6.4 are shown graphically in Fig. 6.8, where critical total pressure recovery is

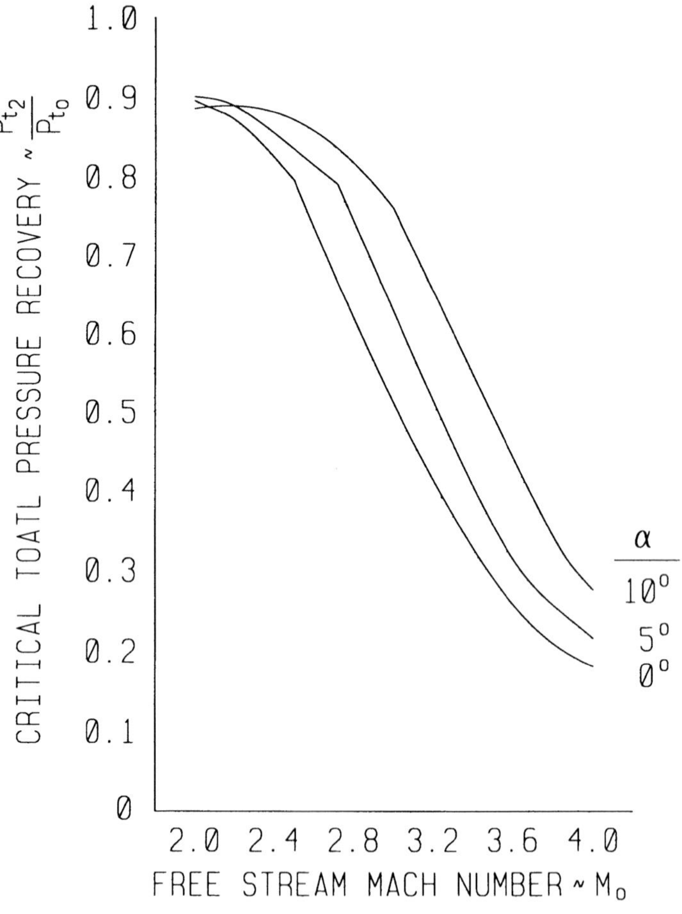

Fig. 6.8 Estimated variation of critical total pressure recovery with freestream Mach number for a shrouded two-dimensional inlet at three angles of attack.

plotted against freestream Mach number at the three angles of attack. The breaks in the slopes of the curves represent the design points and illustrate the shift in design to a higher freestream Mach number with increasing angle of attack. Above design, each curve has the same slope, and the curves would plot as straight parallel lines or semilog paper. Generally, at all freestream Mach numbers, the curves show that the total pressure recovery of a shrouded downward-compressing two-dimensional inlet increases with increasing angle of attack between 0 and 10 deg.

6.2 Capture Area Ratio

As shown in Fig. 5.1, at the design operating condition for a typical ramjet inlet, the cross-sectional area A_0 of the freestream tube captured by the inlet is equal to the cowl area A_C. The example inlets, defined in Chapter 3, have a cowl height equal to cowl width; thus, the capture shape is a square. Although any imaginable capture shape is possible, the square is the easiest to use in the demonstration of estimating procedures; hence, all subsequent discussion in this section is based on a square capture shape.

During critical operation above design Mach number, the ratio of the area of the freestream tube captured by the inlet to the cowl area can be determined from simple geometry to be unity. At critical conditions below design Mach number, where the wedge shock and compression fan have moved upstream from the cowl lip, the inlet capture area ratio can be determined from the continuity equation written between freestream station 0 and the cowl lip plane identified as station 1 in Fig. 6.9. Continuity may be expressed in the form derived in Chapter 1:

$$\frac{Pt_0\left(\dfrac{P}{Pt}\,\mathring{m}\right)_0 A_0}{\sqrt{Tt_0}} = \frac{Pt_1\left(\dfrac{P}{Pt}\,\mathring{m}\right)_1 A_1}{\sqrt{Tt_1}} \tag{6.4}$$

Dividing both sides by the cowl area A_C, recognizing that $Tt_0 = Tt_1$ and

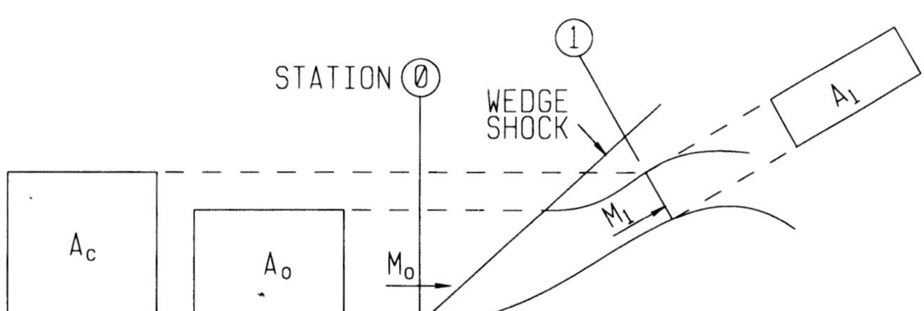

Fig. 6.9 Sketch of inlet operating below design Mach number at critical recovery showing more key parameters for performance-estimating purposes.

rearranging, yields

$$\frac{A_0}{A_C} = \frac{Pt_1}{Pt_0} \frac{\left(\frac{P}{Pt}\mathring{m}\right)_1 A_1}{\left(\frac{P}{Pt}\mathring{m}\right)_0 A_C} \tag{6.5}$$

where A_0/A_C is capture area ratio to be determined, Pt_1/Pt_0 the total pressure recovery across the wedge shock,

$$\left(\frac{P}{Pt}\mathring{m}\right)_1$$

is a function of the Mach number M_1 (given in Ref. 2),

$$\left(\frac{P}{Pt}\mathring{m}\right)_0$$

is a function of the Mach number M_0 (given in Ref. 2), and A_1/A_C the ratio of the physical flow area at the cowl lip plane to the cowl area.

The wedge shock total pressure recovery Pt_1/Pt_0 and the Mach number M_1 at the cowl lip plane, as shown in Fig. 6.9, are the same values determined at corresponding operating conditions in the calculation of overall critical total pressure recovery.

Equation (6.5) has been solved to determine the capture area ratios of the Mach 3.0 design inlet from Chapter 3, and the results are tabulated in Table 6.5.

Design values are given to illustrate that the mathematical procedure gives the same result as simple geometry. Generally, the procedure yields results that correlate with experimental data within the limits of measurement accuracy. However, experimental results can be effected by such factors as the precision with which an inlet is manufactured, the rate of boundary-layer growth along the external compression surface, and ther-

Table 6.5 Calculation of capture area ratio at critical recovery for a Mach 3.0 design example inlet operating at and below the design Mach number

	M_0	M_1	$\frac{Pt_1}{Pt_0}$ ×	$\left(\frac{P}{Pt}\mathring{m}\right)_1$ ×	$\frac{A_1}{A_C}$ ÷	$\left(\frac{P}{Pt}\mathring{m}\right)_0$ =	$\frac{A_0}{A_C}$
Design	3.0	1.99	0.994	0.31776	0.398	0.12558	1.000
	2.5	1.62	0.996	0.41983	↓	0.20168	0.825
	2.0	1.16	0.998	0.52146		0.31512	0.657
		From Table 6.2		Ref. 2	Fig. 3.8	Ref. 2	

mal growth. These factors can shift the shock-on-lip Mach number to a different value (usually higher) than the design point that was originally intended.

A plot of the estimated capture area ratio of the Mach 3.0 design inlet as determined from Eq. (6.5) below design and from geometrical considerations above design over a range of freestream Mach numbers from 2.0 to 4.0 is given in Fig. 6.10 in order to show a graphical representation of a typical capture area variation at critical operating conditions. During supercritical operation, where the terminal normal shock has moved downstream from the throat section, the external flow picture is unchanged, so that the capture area variation with freestream Mach number is identical to that shown in Fig. 6.10. During subcritical operation, where the terminal normal shock has moved upstream from the cowl lip, the inlet spills air and the capture area ratio at any freestream Mach number will be less than that shown in Fig. 6.10. The amount of spillage is a function of the shock standoff distance from the cowl lip. As explained earlier in the discussion of total pressure recovery estimation, no formal estimating procedures are presented in the ramjet literature for subcritical inlet operation.

The capture area characteristics of an inlet operating at angle of attack can best be defined with the aid of the shrouded two-dimensional configuration sketched in Fig. 6.6 and described in the preceding section on total pressure recovery. At and below design Mach number, the capture area ratio can be estimated by the equation of continuity written between stations 0 and 1:

$$\frac{A_0}{A_C} = \frac{Pt_S}{Pt_0}\frac{Pt_1}{Pt_S}\frac{\left(\frac{P}{Pt}\mathring{m}\right)_1}{\left(\frac{P}{Pt}\mathring{m}\right)_0}\frac{A_1}{A_C} \tag{6.6}$$

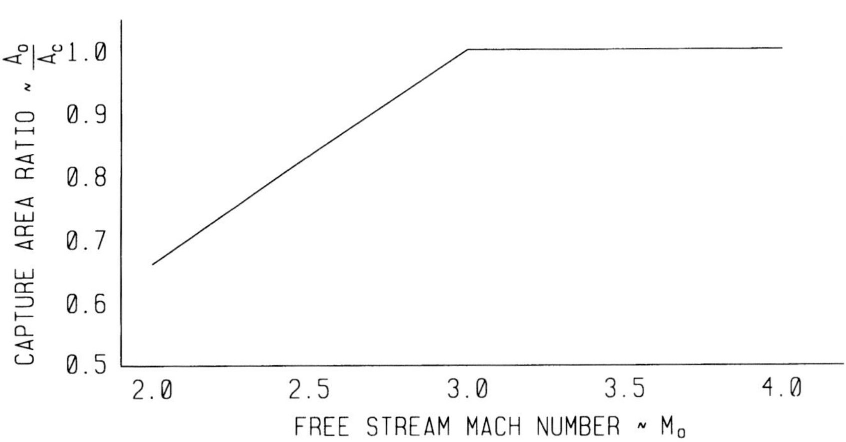

Fig. 6.10 Estimated capture area ratio for a Mach 3.0 design inlet operating at critical conditions over a range of Mach numbers on either side of the design value.

which is the same as Eq. (6.5), except for the shroud shock recovery term Pt_S/Pt_0. At 0-deg angle of attack, the shroud generates only a Mach line; thus, $Pt_S Pt_0 = 1.00$, and Eq. (6.6) reverts to Eq. (6.5). At angles of attack other than 0 deg, the shroud shock recovery is determined as explained earlier in the calculation of total pressure recovery. Similarly, all other terms in Eq. (6.6) may be determined as described in the development of Table 6.5. Above design Mach number, the capture area ratio at any angle of attack can be determined by geometry, and full capture can be achieved only when all shocks (shroud, wedge, compression fan) fall on or inside the cowl lip. Although the shroud has no effect on basic inlet capture at 0-deg angle of attack, the shroud can amplify inlet capture at higher angles of attack at Mach numbers above design, as illustrated in Fig. 6.11. Also shown in Fig. 6.11 are the geometrical relationships used to determine capture area ratios:

$$\frac{A_0}{A_C} = \frac{w\,[z\,\sin(\theta + \alpha)]}{w\,(z\,\sin\theta)} = \frac{\sin(\theta + \alpha)}{\sin\theta} \tag{6.7}$$

Obviously, as the shroud length increases, the angle θ becomes smaller, and the capture area ratio increases, approaching infinity in the extreme.

The procedures described earlier have been applied to the particular basic inlet and shroud combination introduced in the sketches in Fig. 6.7 and shown at three operating angles of attack. Capture area ratio results are tabulated in Table 6.6.

The results of Table 6.6 are shown graphically in Fig. 6.12, where capture area ratio is plotted against freestream Mach number at the three angles of attack. The breaks in the slopes of the curves represent the design points and illustrate the shift in design to a higher freestream Mach number with increasing angle of attack. The second break in the $\alpha = 5$ deg curve represents shroud shock on the lip, which occurs at a higher freestream Mach number than the design (wedge shock and compression fan focal point on lip) condition. At a 10-deg angle of attack, the design condition and shroud shock on the lip occur at nearly the same Mach number. At all Mach numbers above about 2.1, the curves show increasing capture area ratio with increasing angle of attack. At Mach numbers of 3.0 and

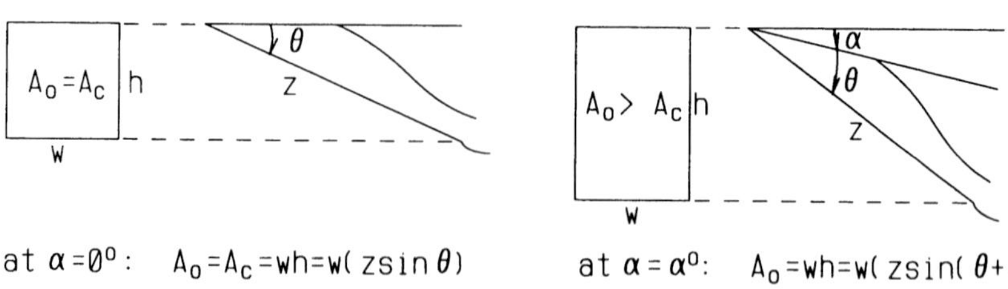

Fig. 6.11 Sketch of shrouded two-dimensional inlet operating at or above design Mach number showing capture amplification at angle of attack and geometrical relationships.

Table 6.6 Calculation of capture area ratio at critical recovery for a shrouded two-dimensional inlet at three angles of attack

M_0	$\frac{Pt_S}{Pt_0}$	×	$\frac{Pt_1}{Pt_S}$	×	$\left(\frac{P}{Pt}\mathring{m}\right)_1$	×	$\frac{A_1}{A_C}$	÷	$\left(\frac{P}{Pt}\mathring{m}\right)_0$	=	$\frac{A_0}{A_C}$
$\alpha = 0$ deg, $M_D = 2.5$ Shroud Mach line on cowl lip at $M_0 = 3.30$											
2.5	1.000		0.995		0.41873		0.481		0.20168		1.000
2.2	↓		0.998		0.48273		↓		0.26523		0.874
2.0			0.999		0.51888				0.31512		0.791
2.5 → 4.0			$\frac{\sin 17.61 \text{ deg}}{\sin 17.61 \text{ deg}} =$								1.000
$\alpha = 5$ deg, $M_D = 2.72$ Shroud shock on lip at $M_0 = 3.08$											
2.72	0.995		0.995		0.41873		0.481		0.16393		1.216
2.5	0.995		0.998		0.46360		↓		0.20168		1.098
2.2	0.999		1.000		0.51875				0.26523		0.940
2.0	1.000		1.000		0.53177				0.31512		0.812
2.72 → 3.08			Straight line between 1.216 and 1.271								
3.08 → 4.00			$\frac{\sin 22.61 \text{ deg}}{\sin 17.61 \text{ deg}} =$								1.271
$\alpha = 10$ deg, $M_D = 3.0$ Shroud shock on lip at $M_0 = 2.98$											
3.0	0.962		0.994		0.41873		0.481		0.12558		1.534
2.8	0.968		0.995		0.45471		↓		0.15193		1.387
2.5	0.975		0.998		0.50625				0.20168		1.175
2.2	0.982		1.000		0.53177				0.26523		0.947
2.0	0.985		1.000		0.53177				0.31512		0.800
2.98 → 4.00			$\frac{\sin 27.61 \text{ deg}}{\sin 17.61 \text{ deg}} =$								1.532

above at 10-deg angle of attack, this particular shrouded inlet is capable of capturing a freestream tube of air with a cross-sectionable area more than 1.5 times the cowl area.

6.3 Drag

The discussion thus far in the text has been devoted almost exclusively to the positive aspects of inlet performance (i.e., the characteristics that contribute to the production of useful engine thrust). The negative side of

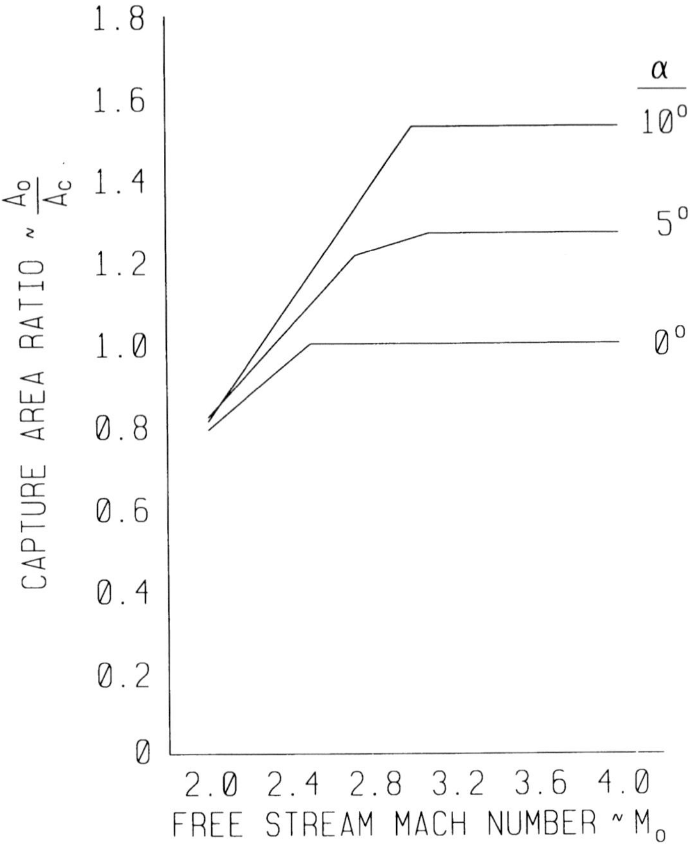

Fig. 6.12 Estimated variation of capture area ratio with freestream Mach number for a shrouded two-dimensional inlet operating at critical conditions at three angles of attack.

inlet performance involves the generation of forces that act to detract from useful thrust. As defined in Chapter 1, useful (also called net) engine thrust is determined from a summation of all of the forces acting on an engine in the axial direction and is shown mathematically:

$$F = \rho_6 A_6 V_6^2 + P_6 A_6 - (\rho_0 A_0 V_0^2 + P_0 A_0) - \int_0^6 [(P_A \sin\theta) + \tau_A] \, dA_w \tag{1.9}$$

where $\rho_6 A_6 V_6^2 + P_6 A_6$ is the stream thrust at station 6; $\rho_0 A_0 V_0^2 + P_0 A_0$ the stream thrust at station 0; $\int_0^6 P_A \sin\theta \, dA_w$ the integral of the pressure forces acting in the drag direction; and $\int_0^6 \tau_A \, dA_w$ the integral of the viscous forces acting on all external surfaces.

In the subsequent discussion in Chapter 1, the integral drag terms were ignored and a net jet thrust coefficient was derived. Although net jet thrust levels are essential in the construction of an engine performance map, final engine selection is usually based on net thrust levels, which are determined from a solution of the integral term in Eq. (1.9).

The drag forces (or the forces acting in an axial direction in opposition to the thrust force) on a flow-through ramjet engine are shown in the partial schematic in Fig. 6.13. These forces may be divided into three categories: pressure P_A, viscous τ_A, and momentum $m_0(V_0 - V_6)$. Pressure forces include the increase in pressure above ambient on all forward-facing surfaces because of shock waves and the decrease in pressure below ambient on all rearward-facing surfaces because of Prandtl-Meyer expansion. Viscous forces embrace skin friction on all wetted surfaces. Momentum forces usually apply to an inlet bleed device where some momentum loss is inevitable. The pressure forces acting on the forward-facing surfaces are subdivided into the additive, or spillage, drag force applied to the capture stream tube boundary during inlet operation below design Mach number; the cowl lip wave drag; and the wave drag of the rounded, or tapered, cowl section between the downstream end of the cowl lip and the full engine diameter section. These drag forces will be treated in detail subsequently. The pressure forces acting on the rearward-facing surfaces involve suction on boattail and base regions of the engine. Normally, these forces are not applied to an inlet, although an offset inlet with an external fairing would experience a boattail drag. Under any installation circumstances, the drag would be determined by the methods presented in Ref. 21, and no estimation techniques are included herein. Skin-friction drag is usually estimated by the external aerodynamicist for the entire vehicle, including the engine sections, but it is useful for the inlet designer to have knowledge of the principal elements of the estimating technique. Last, methods to determine inlet bleed flow rate and bleed drag are presented. Drag forces associated with a particular installation geometry (e.g., vehicle boundary-layer diverters) will be discussed in the next chapter.

Drag forces may be nondimensionalized by specifying a drag coefficient that has the form

$$C_D = \frac{D}{q_0 A_{\text{ref}}} \tag{6.8}$$

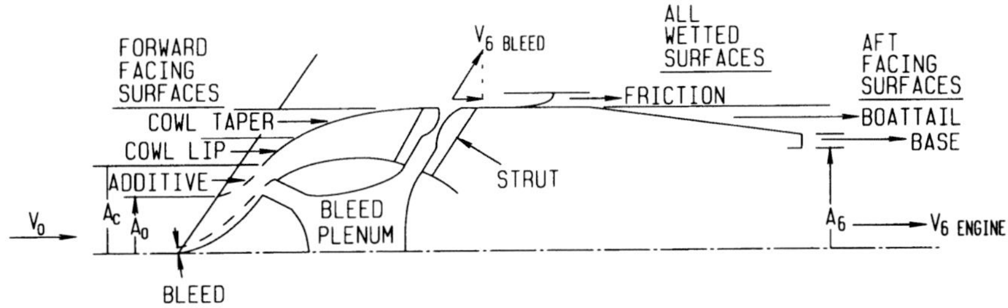

Fig. 6.13 Partial schematic of a flow-through ramjet engine showing forces acting to produce drag.

where D is drag; q_0 freestream dynamic pressure; $= \frac{1}{2}\rho_0 V_0^2 = (\gamma/2)P_0 M_0^2$; and A_{ref} an arbitrarily chosen reference area and would be the same area selected for C_F as defined in Eq. (1.14).

If each of the forward-facing surface pressures in Fig. 6.13 were referenced to the ambient value P_0, the integral of the engine pressure drag forces would be written as

$$D = (P_{\text{add}} - P_0)\,\mathrm{d}A_{\text{add}} + (P_{CL} - P_0)\,\mathrm{d}A_{CL} + (P_{\text{taper}} - P_0)\,\mathrm{d}A_{\text{taper}} \tag{6.9}$$

and the corresponding drag coefficient,

$$C_D = \frac{(P_{\text{add}} - P_0)\,\mathrm{d}A_{\text{add}}}{q_0\,A_{\text{ref}}} + \frac{(P_{xx} - P_0)\,\mathrm{d}A_{xx}}{q_0\,A_{\text{ref}}} \tag{6.10}$$

where P_{xx} is the surface pressure on any forward-facing element of the engine with a projected frontal area of $\mathrm{d}A_{xx}$, but Ref. 1 defines a pressure coefficient,

$$C_p = \frac{(P_{\text{surface}} - P_0)}{q_0} \tag{6.11}$$

Hence, the drag coefficient for any segment of the forward-facing surfaces can be written as

$$C_{D_P} = C_p \frac{\mathrm{d}A}{A_{\text{ref}}} \tag{6.12}$$

Curves of pressure coefficient C_p vs flow deflection angle over a range of freestream Mach number M_0 from 1.05 to ∞ are given in Ref. 1 for both two-dimensional and conical flow.

Equation (6.12) may be used to estimate the pressure drag of each of the forward-facing inlet surfaces, but the procedures used in the solution are different for each surface element. Although the estimating accuracy for each element is unknown, the total drag estimated from the summation of all of the elemental drags (both pressure and friction) for two different engine drag models correlated with drag balance results within experimental accuracy. The procedures used in the estimates of the elemental pressure drag coefficients are as follows:

1) The cowl lip is always treated as a wedge whether the inlet is two-dimensional or axisymmetric. The tangent wedge approach is used; i.e., the flow is turned from the freestream Mach number through a deflection angle equal to the external cowl lip angle, as shown in Fig. 6.14. If the external cowl lip angle is steep enough to cause shock-wave detachment at lower supersonic freestream Mach numbers, the flow is assumed to be compressed isentropically to a wedge surface Mach number of 0.90, and

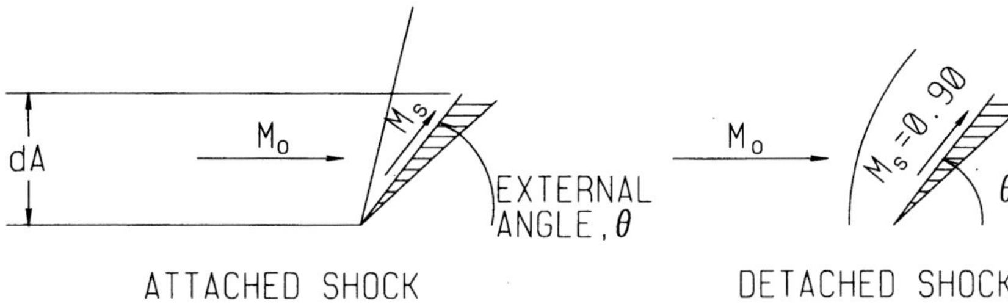

Fig. 6.14 Cowl lip details for drag estimation.

the pressure coefficient C_P must be calculated rather than found directly from the curves of Ref. 1. A detached shock-wave picture is also shown in Fig. 6.14. The projected frontal area dA of the cowl lip and the other forward-facing surfaces can best be determined from a large-scale drawing of the inlet, where the boundaries of each elemental surface may be easily discerned.

2) The tapered, or curved, section of the cowl is approximated by a series of flat surfaces as shown in Fig. 6.15. Each separate surface is treated as a tangent wedge, and, normally, the deflection angles are low enough that shock-wave detachment does not occur within the freestream Mach number range of interest.

3) During inlet operation below the design Mach number (Fig. 6.16), the boundary of the captured stream tube acts as an extension of the cowl and experiences pressure forces that contribute to the total drag. Inasmuch as the boundary is curved by the isentropic surface compression fan, the Mach number and pressure are constantly changing along the length of the boundary from the wedge shock to the cowl lip. Instead of a time-consuming pressure-area integration, the curved boundary is assumed to be adequately represented as the straight-line hypotenuse of a right triangle with a centroid one-third of the height from the base at the cowl lip plane, as shown in Fig. 6.16, with the average pressure $\overline{P}$ acting at the centroid. The average pressure is commensurate with compression of the flow through two-thirds of the total turning angle specified in the design of the supersonic diffuser. Below the Mach number where the cowl lip shock detaches there is no analytical solution to the additive drag coefficient.

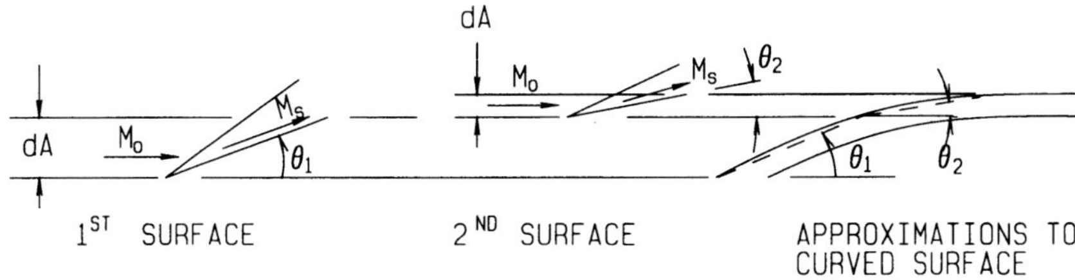

Fig. 6.15 Tapered cowl section details for drag estimation.

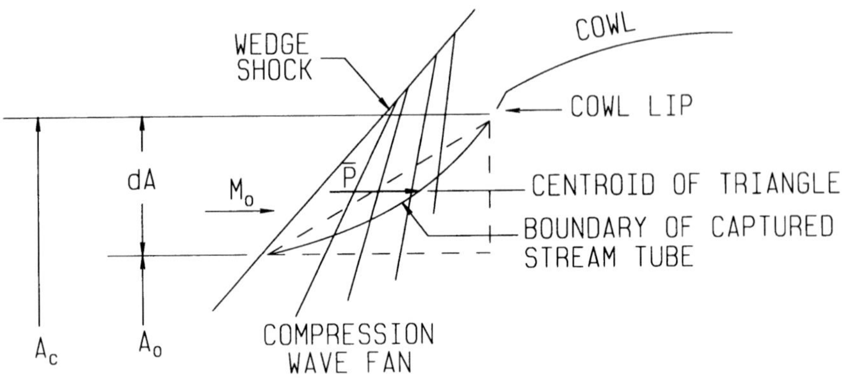

Fig. 6.16 Additive details for drag estimation.

The application of Eq. (6.12) to each of the forward-facing elements of the Mach 3.0 design example inlet of Chapter 3 with the procedural rules described earlier gives the drag coefficient results given in Table 6.7.

Ratios of the projected frontal area dA of each elemental surface to the cowl area A_C were determined from measurements on an inlet layout that was considerably larger than the inlet sketch drawn in Fig. 3.8. The cowl area was chosen for reference simply as a convenience, and the drag coefficients given in Table 6.7 can be referenced to any other desired area by multiplying the results by the ratio of A_C to the new reference area, A_{ref}. It should be noted that any drag coefficient values are meaningless unless the corresponding reference area is identified. The tabulated results of Table 6.7 are shown graphically in Fig. 6.17, where the pressure drag coefficient buildup is plotted against the freestream Mach number. The break in the slope of the total pressure drag coefficient (topmost) curve occurs at the inlet design Mach number. Above design, the drag coefficient decreases slowly with increasing freestream Mach number. Below design, the drag coefficient increases very rapidly with decreasing freestream Mach number. Most of the rise is caused by the additive term, and the levels suggest that inlet spillage can be expensive in terms of useful thrust. The dashed part of the total drag coefficient curve is an approximation based on experimental data gathered in tests of other all-external compression inlets. The dashed part of the curve identifies the Mach number range where the cowl lip shock is detached, and this condition produces the characteristic hump in the cowl lip wave drag coefficient curve. Typically, the pressure drag coefficient levels will decrease steadily through the transonic Mach number region.

The nondimensionalized viscous drag coefficient has the form

$$C_{D_F} = C_f \frac{\mathrm{d}A_w}{A_{\mathrm{ref}}} \tag{6.13}$$

where C_f is a friction coefficient, and dA_w the wetted surface area of inlet or engine or vehicle.

Table 6.7 Calculation of pressure drag coefficients for a Mach 3.0 design example inlet operating over a wide range of freestream Mach numbers

Cowl lip wave drag coefficient
deflection angle = 28.5 deg $\qquad$ $dA/A_C = 0.072$

$A_C = A_{ref}$

M_0	C_P	$\times\ \dfrac{dA}{A_C}$	$= C_{D_{CL}}$
4.0	0.67	0.072	0.048
3.5	0.71	│	0.051
3.0	0.78	│	0.056
2.5	0.93	│	0.067
2.4	1.01	↓	0.073

Shock detached		$\dfrac{P_S}{Pt_S}$	$-\ \dfrac{P_0}{Pt_0}$	$\div\ \dfrac{q_0}{Pt_0}$	$= C_P$	$\times\ \dfrac{dA}{A_C}$	$= C_{D_{CL}}$
│	2.2	0.5913	0.0935	0.3169	1.571	0.072	0.113
│	2.0	│	0.1278	0.3579	1.295	│	0.093
↓	1.8	↓	0.1740	0.3947	1.057	↓	0.076

Cowl taper drag coefficient

1st SURFACE ∢ = 15 deg $\dfrac{dA}{A_C} = 0.089$ $\qquad$ 2nd SURFACE ∢ = 6 deg $\dfrac{dA}{A_C} = 0.027$

M_0	C_P	$\times\ \dfrac{dA}{A_C}$	$= C_{D_{t1}}$	M_0	C_P	$\times\ \dfrac{dA}{A_C}$	$= C_{D_{t2}}$
4.0	0.24	0.089	0.022	4.0	0.07	0.027	0.002
3.5	0.26	│	0.023	3.5	0.08	│	0.002
3.0	0.29	│	0.026	3.0	0.09	│	0.002
2.5	0.33	│	0.030	2.5	0.11	│	0.003
2.4	0.35	│	0.031	2.4	0.11	│	0.003
2.2	0.38	│	0.034	2.2	0.12	│	0.003
2.0	0.43	│	0.038	2.0	0.14	│	0.004
1.8	0.50	↓	0.045	1.8	0.16	↓	0.004

Additive drag coefficient (below $M_D = 3.0$)
Average $\overline{C}_P$ based on flow turning through two-thirds of 24 deg = 16 deg

$$\frac{dA}{A_C} = \frac{A_C - A_0}{A_C} = 1 - \frac{A_0}{A_C}$$

M_0	ν_0	ν_0 − 16 deg	$\overline{M}$	$\dfrac{\overline{P}}{Pt}$	$-\ \dfrac{P_0}{Pt_0}$	$\div\ \dfrac{q_0}{Pt_0}$	$= \overline{C}_P$	$\times \left(1 - \dfrac{A_0}{A_C}\right)$	$= C_DA$
2.9	47.79	31.79	2.20	0.09352	0.03165	0.1873	0.330	0.030	0.010
2.7	43.62	27.62	2.045	0.1192	0.04295	0.2192	0.348	0.110	0.038
2.5	39.12	23.12	1.88	0.1539	0.05853	0.2561	0.372	0.175	0.065
2.4	36.75	20.75	1.80	0.1740	0.06840	0.2758	0.383	0.260	0.100
2.2	Cowl lip shock detachment:		No analytical solution						
2.0									
1.8									

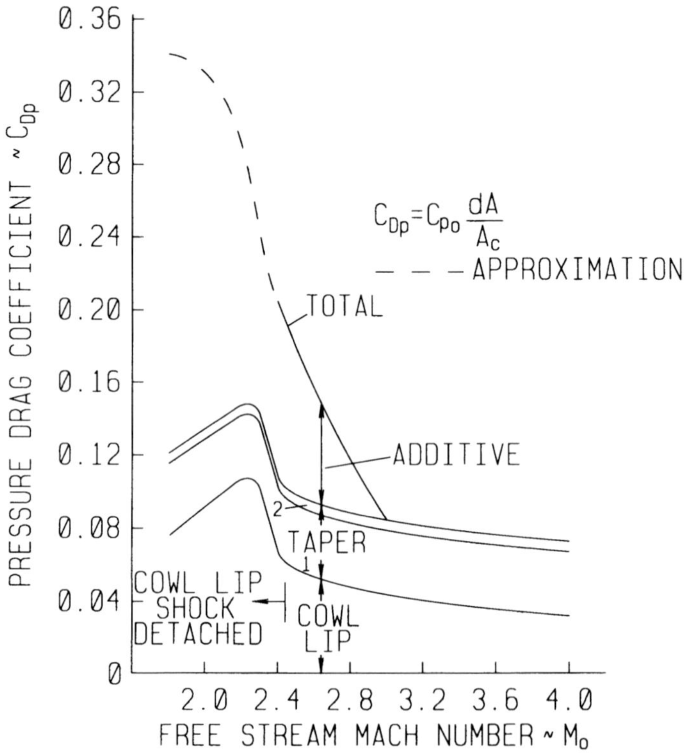

Fig. 6.17 Estimated variation of the pressure drag coefficient with freestream Mach number for the forward-facing surfaces of a Mach 3.0 design inlet.

By definition,

$$C_f = \frac{\tau_0}{\frac{1}{2}\rho V^2}$$

and

$$\tau_0 = \mu \frac{\mathrm{d}u}{\mathrm{d}y} \text{ at } y = 0$$

(See Chapter 4.)

Values of C_f for flow along a flat-plate with specific Reynolds number and Mach number limitations have been determined:

1) very-low-Reynolds-number laminar flow with parabolic velocity distribution [the exact solution of Blasius (1908)]:

Average skin friction:

$$C_f = \frac{1.327}{\sqrt{Re}} \tag{4.14}$$

2) turbulent flow with 1/7 root profile (valid to $Re = 3 \times 10^7$, but results 14.5% lower than experimental date at $M_0 = 2.5$):

Von Kármán (1934):

$$C_f = \frac{0.074}{(Re)^{1/5}} \tag{4.17}$$

3) turbulent (incompressible) flow with any profile (Re range of 10^6–10^9):

Schlicting (1949):

$$C_f = \frac{0.472}{(\log_{10} Re)^{2.58}}$$

4) compressible turbulent flow with any profile (agrees with experimental data within ±2% at M_0 from $0 \rightarrow 4.0$):

Extended Frankl-Voishel[22]:

$$C_f = \frac{0.472}{(\log_{10} Re)^{2.58}\left(1 + \dfrac{\gamma - 1}{2} M_0^2\right)^{0.467}} \tag{6.14}$$

where $Re = \rho_0 U_0 \ell / \mu_0$ is based on freestream conditions and length ℓ measured from the leading edge of the wetted surface, and $\gamma = 1.4$.

The solution to Eq. (6.14) is shown in Fig. 6.18, where the average skin-friction coefficient C_f is plotted against Reynolds number Re as a function of freestream Mach number M_0. It is evident that increasing Mach number at any Reynolds number decreases skin friction, and at any constant Mach number, skin friction decreases as Reynolds number increases. In the latter case increases in Re must be accomplished by decreases in operating altitude. Obviously, high-speed flight at low altitude will generate the least amount of friction drag, and the converse is true. For practical missile applications selected Mach number altitude points on the flight trajectory are used to calculate key flight Reynolds numbers and enter Fig. 6.18 to find the corresponding average skin-friction coefficients. Equation (6.13) is then solved for the friction drag coefficients, which are added to the pressure drag coefficients determined from Eq. (6.12) to form the complete drag picture (for an inlet without bleed). Estimates and a few drag balance measurements correlate quite well, except that the friction drag for cylindrical bodies is about 20% higher than the values obtained with Eq. (6.14).

If some of the air captured by an inlet is diverted through passages separate from the main engine, then that air will have a momentum history that is different from the main airflow. The ideal bleed arrangement is one where the bleed air is diverted into a plenum chamber with no loss in total pressure and expanded through an exit nozzle in an axial direction. A sketch of a bleed arrangement is given in Fig. 6.19, which shows the prin-

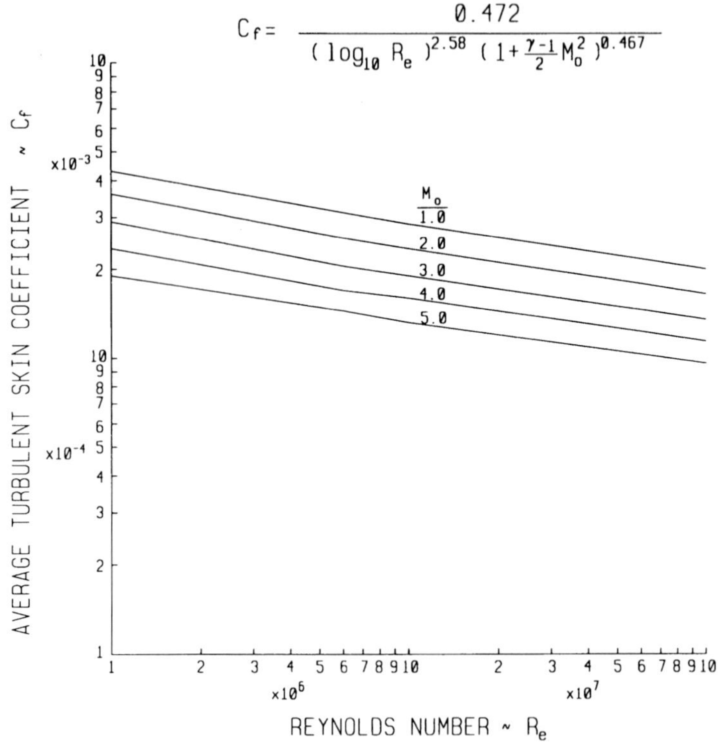

Fig. 6.18 Variation of friction coefficient with Reynolds number for several free-stream Mach numbers.

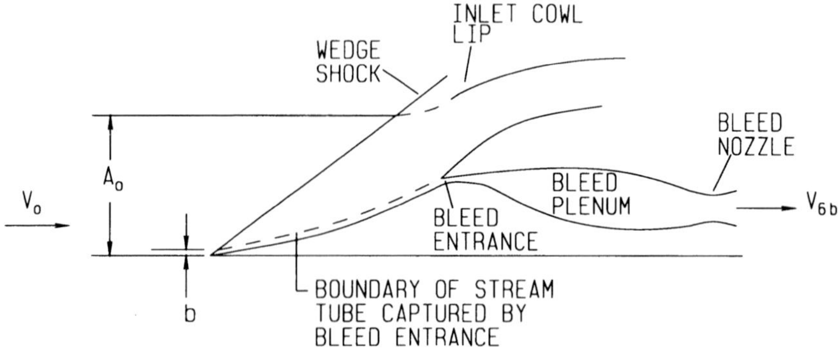

Fig. 6.19 Sketch of a bleed arrangement showing key parameters for estimating drag.

cipal parts together with some of the key parameters necessary to determine the momentum change between entrance and exit:

Bleed drag = inlet momentum − exit momentum

$$D = b\rho_0 A_0 V_0 V_0 - b\rho_0 A_0 V_0 V_6 \tag{6.15}$$

Defining

$$C_{D_B} = \frac{D}{\frac{1}{2}\rho_0 V_0^2 A_C} \tag{6.16}$$

Then

$$C_{D_B} = \frac{b\rho_0 A_0 V_0 V_0}{\frac{1}{2}\rho_0 V_0^2 A_C} - \frac{b\rho_0 A_0 V_0 V_6}{\frac{1}{2}\rho_0 V_0^2 A_C}$$

or

$$C_{D_B} = 2b\frac{A_0}{A_C}\left[1 - \frac{V_6}{V_0}\right] \tag{6.17}$$

If the bleed flow is dumped overboard in the tangential direction (no axial component), then $V_6 = 0$, and the expression becomes

$$C_{D_B} = 2b\frac{A_0}{A_C} \tag{6.18}$$

This is the usual case for side-wall bleed with flush holes, where it is impractical to attempt to recover any momentum.

Defining a velocity coefficient for the bleed exit nozzle,

$$C_v = \frac{V_6}{V_{6_{\text{ideal}}}} \tag{6.19}$$

and substituting in Eq. (6.17) yields

$$C_{D_B} = 2b\frac{A_0}{A_C}\left[1 - C_v\frac{V_{6_{\text{ideal}}}}{V_0}\right] \tag{6.20}$$

but, by definition, the kinetic energy efficiency,

$$n_{kE} = \frac{{V_{6_{\text{ideal}}}}^2}{V_0^2} \tag{6.21}$$

The variable n_{kE} will be derived in the next section.

Substituting Eq. (6.21) into Eq. (6.20),

$$C_{D_B} = 2b\frac{A_0}{A_C}\left[1 - C_v\sqrt{n_{kE}}\right] \tag{6.22}$$

For a well-designed nozzle, $C_v = 0.98$; thus, finally, the bleed system drag coefficient for general use is

$$C_{D_B} = 2b\frac{A_0}{A_C}(1 - 0.98\sqrt{n_{kE}}) \tag{6.23}$$

If the bleed system recovery is very low, n_{kE} becomes a small number, and

$$C_{D_B} = 2b\frac{A_0}{A_C} \tag{6.18}$$

The result is the same as dumping the bleed overboard with no attempt to recover any momentum.

If the bleed system recovery is 100%, n_{kE} becomes 1.0, and

$$C_{D_B} = 0.04b\frac{A_0}{A_C} \tag{6.24}$$

This is the lowest drag situation possible with a given bleed flow rate.

The drag for any bleed arrangement should lie between the extremes represented by Eqs. (6.18) and (6.24) and, realistically, be much closer to the high levels calculated in Eq. (6.18).

The bleed flow rate, or bleed mass fraction, is determined from continuity:

$$b = \frac{Pt_b\left(\frac{P}{Pt}\mathring{m}\right)_b\frac{A_b}{A_C}}{Pt_0\left(\frac{P}{Pt}\mathring{m}\right)_0\frac{A_0}{A_C}} \tag{6.25}$$

where the bleed flow conditions, given by the subscript b, can be determined at any station in the bleed arrangement (i.e., at the entrance, in the plenum, or at the nozzle throat).

As a first step in the illustration of the procedures used in solving the bleed flow rate equation and the bleed drag coefficient equation, it is postulated that the familiar example inlets of Chapter 3 are equipped with bleed devices. Two cases, which are representative of the opposite ends of the bleed drag spectrum, are examined.

Table 6.8 Calculation of bleed flow rates for three example inlets with sidewall holes

M_0	M_1	$\frac{P_1}{Pt_1}$	×	$\frac{Pt_1}{Pt_0}$	×	$\left(\frac{P}{Pt}\mathring{m}\right)_b$	×	C	×	$\frac{A_b}{A_C}$	÷	$\left(\frac{P}{Pt}\mathring{m}\right)_0$	×	$\frac{A_0}{A_C}$	=	b
2.0	1.47	0.2845		1.000		0.53177		0.5		0.083		0.3151		1.00		0.020
3.0	1.99	0.1298		0.994		↓		↓		↓		0.1256		↓		0.023
4.0	2.52	0.0567		0.990		↓		↓		↓		0.0496		↓		0.025
Table 6.1		Ref. 2		Table 6.1		Ref. 2 $M_b = 1.0$		Assumed or selected				Ref. 2				

First case

Both side walls of each inlet are drilled in the pattern shown in Fig. 6.20 to produce a total hole area to cowl area ratio of 0.083, which was calculated to yield about a 2% bleed flow rate for the Mach 2.0 design inlet. The bleed flow is dumped directly overboard (no plenum or nozzle) at right angles to the freestream, so that there is no momentum recovery. The driving pressure of the bleed flow is the potential flow static pressure at station 1 in the inlets. The holes are assumed to be choked and have a flow coefficient equal to 0.5.

The solution to Eq. (6.25) is given in Table 6.8. The corresponding drag coefficient from Eq. (6.18) is given in Table 6.9.

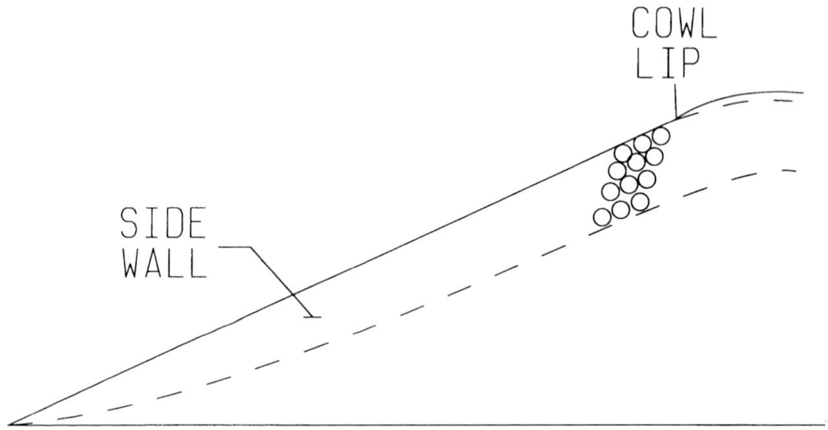

Fig. 6.20 Pattern of flush bleed holes drilled in side walls of three two-dimensional example inlets.

Table 6.9 Calculation of bleed drag coefficients for three example inlets with sidewall holes.

M_0	2	×	b	×	A_0/A_C	=	C_{DB}
2.0	↓		0.020		1.00		0.040
3.0			0.023		↓		0.046
4.0			0.025				0.050

Second case

Each inlet has a ram slot, or scoop, located as shown in Fig. 6.21 with a height on the order of the compression surface boundary-layer thickness. Total slot flow area to cowl area ratio is 0.02. The driving pressure of the bleed flow is the total pressure at station 1, and the flow coefficient of the slot is 1.00. The only loss through the bleed arrangement is caused by a

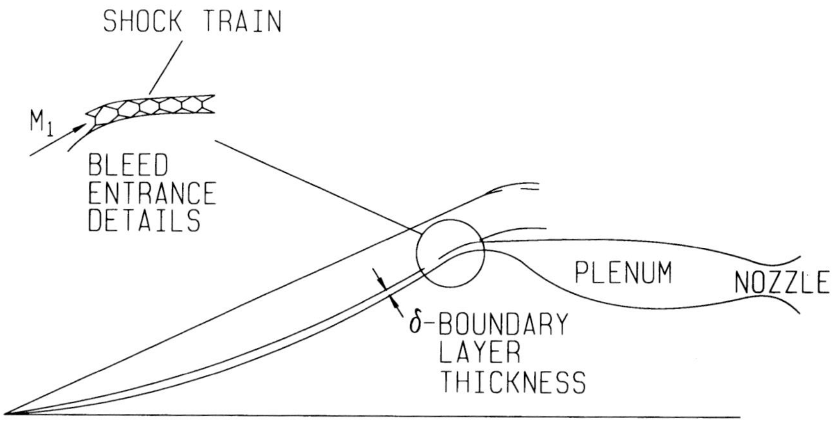

Fig. 6.21 Sketch of ram slot for bleed on compression surfaces of three two-dimensional example inlets.

normal shock at Mach number M_1 and the flow is expanded through an exit nozzle in the axial direction.

The solution to Eq. (6.25) is seen in Table 6.10. The corresponding drag coefficient from Eq. (6.23) is given in Table 6.11.

If the bleed flow rates calculated in case 2 could have been processed with no total pressure loss ($n_{kE} = 1.00$), the drag coefficients would have been determined from Eq. (6.24) (see Table 6.12).

Both the bleed flow rate and bleed drag coefficient results for the two cases are plotted against freestream Mach number in Fig. 6.22. Although the flow values for each case and the drag values for each case are connected, each curve represents a locus of probable design point performance for an infinite number of inlets designed in the Mach number range of 2.0–4.0.

An example of off-design performance for a single inlet is shown by the curves for the shrouded two-dimensional inlet, which was used in the demonstration of angle-of-attack performance (pressure recovery and capture area) estimation. These latter curves result from calculations based on measurements of bleed flow through an instrumented passage. The bleed entrance is a flush slot located in the throat region. Although the terminal normal shock did not influence bleed driving pressure, a misaligned cowl lip generated a shock that increased the driving pressure at the lower Mach numbers with a maximum effect at the inlet design Mach number of 2.5.

It is apparent from Fig. 6.22 that the magnitude of the bleed drag coefficient cannot be deduced from the magnitude of the bleed flow rate. In fact, the lowest flow rate is commensurate with the highest drag level, and vice versa. The bleed flow rate does directly effect the bleed drag level as shown by the equations, but the total pressure recovery of the bleed plays a greater role in determining ultimate drag level. In most cases of practical interest, low bleed flow rates, on the order of 1 or 2%, are usually indicative of flush holes dumping the bleed flow with little or no momentum recovery,

Table 6.10 Calculation of bleed flow rates for three example inlets with compression surface ram slot

M_0	M_1	$\frac{P_{1'}}{Pt_1}$	×	$\frac{Pt_1}{Pt_0}$	×	$\left(\frac{P}{Pt}\mathring{m}\right)_b$	×	$\frac{A_b}{A_C}$	÷	$\left(\frac{P}{Pt}\mathring{m}\right)_0$	×	$\frac{A_0}{A_C}$	=	b
2.0	1.47	0.939		1.000		0.53177		0.02		0.31512		1.00		0.032
3.0	1.99	0.726		0.994		↓		↓		0.1259		↓		0.061
4.0	2.52	0.491		0.990						0.0496				0.104
Table 6.1		Ref. 2		Table 6.1		Ref. 2 $M_{b=1.0}$		Selected		Ref. 2				

Table 6.11 Calculation of bleed drag coefficients for three example inlets with compression ram slot

M_0	2	×	b	×	A_0/A_C	$[1 - 0.98\sqrt{n_{kE}}]$	=	C_{DB}
2.0	↓		0.032	↓	1.00	0.989		0.002
3.0			0.061			0.973		0.006
4.0			0.104			0.963		0.012

whereas higher bleed flow rates (e.g., 5–10%) are commensurate with bleed arrangements where pressure recoveries of various levels are possible depending on the geometrical constraints imposed on the configuration. Although a bleed total pressure recovery of unity is impossible to achieve, the curve defined by Eq. (6.24) does provide a goal and a reference point. It is not known whether the total pressure recoveries (n_{kE}) postulated in the second case are attainable in practice. The n_{kE} based on measurements with the Mach 2.5 design shrouded two-dimensional inlet ranged from a high of 0.38 at $M_0 = 2.16$ to a low of 0.16 at $M_0 = 2.95$.

6.4 Efficiency

Both the positive and negative aspects of inlet design, construction, and performance have now been discussed, but each aspect has been treated as a separate entity with only passing attention to interactions between good and bad. In Chapter 2 supersonic diffusers were rated according to performance capability, and it was indicated that there is a correlation between increasing total pressure recovery, increasing design complexity, and increasing drag. Of five configurations examined, the middle one in Fig. 2.15 (5-deg wedge plus isentropic) was stipulated to yield a balanced compromise between pressure recovery and drag with middling design complexity. This section of the text has been assembled to prove the stipulation and provide a procedure whereby tradeoffs between good and bad inlet characteristics can be weighed, and workable compromises can be selected on a rational analytical basis.

The procedure begins with a method for measuring the relative value of the total pressure recovery capability of a family of inlets designed at the same level of complexity but at different design Mach numbers. For example, the three conservatively designed inlets, which were constructed in

Table 6.12 Calculation of bleed drag coefficients for three example inlets with compression surface ram slot and no bleed system total pressure loss

M_0	0.04	×	b	×	A_0/A_C	=	c_{DB}
2.0	↓		0.032		1.00 ↓		0.0013
3.0			0.061				0.0024
4.0			0.104				0.0042

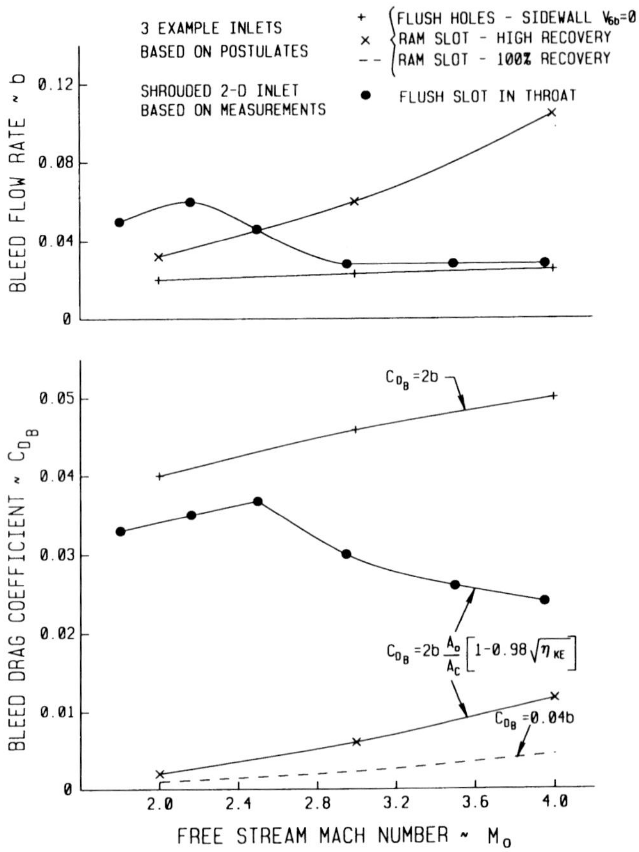

Fig. 6.22 Estimated variation of bleed flow rate and bleed drag coefficient with freestream Mach number for several inlets.

Chapter 3 and subsequently used as representative configurations to illustrate the application of performance-estimating procedures, encompass a wide range of critical total pressure recovery levels: 0.845 at Mach 2.0 through 0.650 at Mach 3.0 to 0.438 at Mach 4.0. A direct comparison of these numbers would rate the Mach 2.0 inlet almost twice as capable as the Mach 4.0 inlet. However, inasmuch as the Mach 4.0 inlet is operating at a kinetic energy level on the order of four times greater than the Mach 2.0 inlet, a true comparison of the pressure recovery capabilities of the two inlets should be determined by some process efficiency. The Brayton cycle presented in Chapter 1 describes the inlet process in terms of kinetic energy; thus, a kinetic energy efficiency scale for comparing inlets of any family, design complexity, or design Mach number seems appropriate. The relationship between kinetic energy efficiency n_{kE} and total pressure recovery Pt_2/Pt_0 is derived later in the chapter, and the basic parameters and stations are shown in the sketch in Fig. 6.23. The key to the derivation is

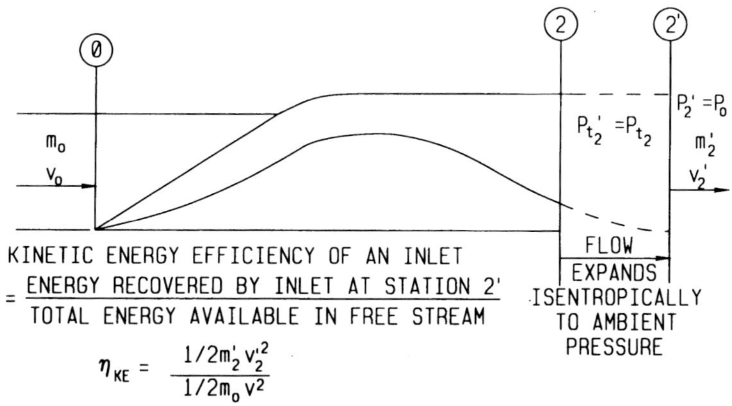

Fig. 6.23 Inlet sketch showing basics involved in kinetic energy efficiency calculation.

that the real flow at the subsonic diffuser exit station 2 is expanded isentropically to a fictitious station 2′, where the static pressure $Ps_{2'}$ equals freestream ambient pressure P_0. With no bleed or leakage between stations, the conservation of mass requires that $m_0 = m_2 = m_{2'}$:

$$n_{kE} = \frac{\frac{1}{2}\, m_{2'}\, V_{2}^2}{\frac{1}{2}\, m_0\, V_0^2} = \frac{V_{2'}^2}{V_0^2} \tag{6.26}$$

Using the relationships shown in Chapter 1 between velocity, speed of sound, and Mach number (i.e., $V = aM$ and $a = \sqrt{(\gamma RT)}$, the equation becomes

$$n_{kE} = \frac{(M_{2'})^2\, T_{2'}}{(M_0)^2\, T_0} \tag{6.27}$$

Dividing the numerator and demoninator by Tt_0, remembering that $Tt_2 = Tt_0$, and recalling the compressible Bernoulli relationships from Chapter 1, i.e.,

$$\left(\frac{Pt}{P}\right)^{\gamma - 1/\gamma} = \frac{Tt}{T} = \left[\frac{\gamma - 1}{2}M^2 + 1\right]$$

the equation becomes

$$n_{kE} = \frac{(M_{2'})^2 \dfrac{T_{2'}}{Tt_0}}{(M_0)^2 \dfrac{T_0}{Tt_0}} = \frac{(M_{2'})^2 \left(\dfrac{Pt_0}{P_0}\right)^{\gamma - 1/\gamma}}{(M_0)^2 \left(\dfrac{Pt_{2'}}{P_{2'}}\right)^{\gamma - 1/\gamma}} \tag{6.28}$$

Because $P_{2'} = P_0$ and $Pt_{2'} = Pt_2$,

$$n_{kE} = \frac{(M_{2'})^2}{(M_0)^2}\left(\frac{Pt_0}{Pt_2}\right)^{\gamma - 1/\gamma} \tag{6.29}$$

From the Bernoulli relationships, $M_{2'}$ may be written

$$(M_{2'})^2 = \frac{2}{\gamma - 1}\left[\left(\frac{Pt_2}{P_0}\right)^{\gamma - 1/\gamma} - 1\right] \tag{6.30}$$

Substituting Eq. (6.30) into Eq. (6.29):

$$n_{kE} = \frac{2}{(\gamma - 1)\, M_0^2}\left[\left(\frac{Pt_2}{P_0}\right)^{\gamma - 1/\gamma} - 1\right]\left(\frac{Pt_0}{Pt_2}\right)^{\gamma - 1/\gamma} \tag{6.31}$$

Performing the multiplication,

$$n_{kE} = \frac{2}{(\gamma - 1)\, M_0^2}\left[\left(\frac{Pt_0}{P_0}\right)^{\gamma - 1/\gamma} - \left(\frac{Pt_0}{Pt_2}\right)^{\gamma - 1/\gamma}\right] \tag{6.32}$$

Substituting

$$\left[1 + \frac{\gamma - 1}{2} M_0^2\right]$$

for

$$\left(\frac{Pt_0}{P_0}\right)^{\gamma - 1/\gamma}$$

and combining terms yields, finally,

$$n_{kE} = 1 - \frac{2}{(\gamma - 1)\, M_0^2}\left[\left(\frac{Pt_0}{Pt_2}\right)^{\gamma - 1/\gamma} - 1\right] \tag{6.33}$$

For $\gamma = 1.4$,

$$n_{kE} = 1 - \frac{5}{M_0^2}\left[\left(\frac{Pt_0}{Pt_2}\right)^{1/3.5} - 1\right] \tag{6.34}$$

The relationship between inlet kinetic energy efficiency and total pressure recovery over a range of freestream Mach numbers from 2.0 to 4.0

is shown in Fig. 6.24. The lines of constant kinetic energy efficiency between a low of 0.88 and a high of 0.96 embrace a region of primary interest for practical inlets with acceptable performance in terms of modern ramjet engine thrust capability. The five dashed lines, which are grouped within the acceptable n_{kE} band at Mach 2.0 and fan out with increasing Mach number, are directly related to the five supersonic diffuser performance curves shown in Fig. 2.15. Each dashed line represents the locus of peak performance for a series of inlets of comparable complexity designed at innumerable Mach numbers within the range of 2.0 to 4.0. The normal shock configuration, the least complex, barely passes the acceptable n_{kE} standard at Mach 2.0 and is not a viable candidate at any Mach number above 2.0. The single-surface 15-deg wedge configuration with one level of complexity above the normal shock has a more shallow slope and is barely outside the acceptable region at Mach 3.0. The middle curve, the isentropic surface with a 5-deg initial wedge, which was termed the conservatively constrained design in Chapter 2, generally lies in the upper half of the acceptable band between n_{kE} levels of 0.92 and 0.94. Most fixed-geometry, all-external compression inlets for practical missile applications would probably have a critical total pressure recovery at the design Mach number on, or very near, this curve. As examples, the measured critical total pressure recovery values of the BOMARC axisymmetric inlet (M_D = 2.6) and of the shrouded two-dimensional inlet (M_D = 2.5), used extensively herein to demonstrate a performance estimation at angle of attack, are plotted at the respective design Mach numbers, and do indeed fall on the middle curve. The fourth curve, commensurate with the fixed-geometry inlet operating at the external compressive turning limit or a variable-geometry inlet operating at the internal running contraction limit, has a

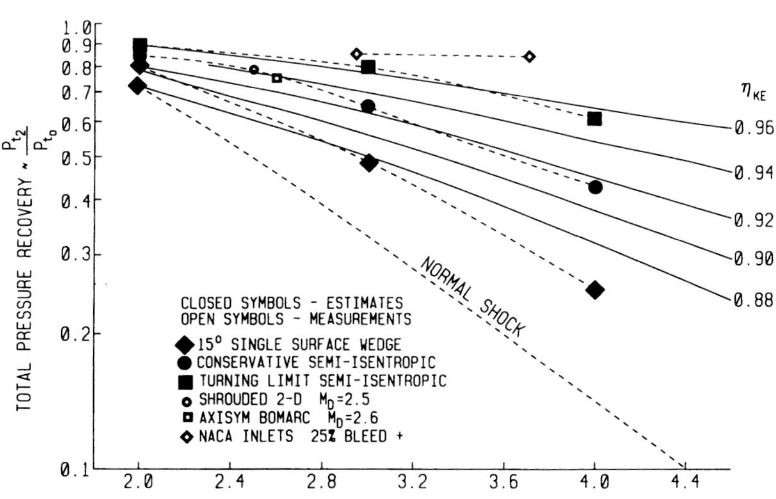

Fig. 6.24 Relationship between kinetic energy efficiency and total pressure recovery over a range of freestream Mach numbers.

slope very close to the $n_{kE} = 0.96$ line, i.e., the upper boundary of the acceptable range for practical inlets. Both of these inlet types will be discussed in subsequent sections of this chapter. The fifth curve connects the measured on-design performance of two variable-geometry NACA inlets[10,12] with all-internal contraction and massive (more than 25%) bleed flow rates. These inlets were designed for research purposes only and indicate that very high n_{kE} levels are achievable with special preparation and carefully controlled test conditions.

A series of three-dimensional inlets featuring double-scoop supersonic diffusers with elliptical capture shapes and round throats were built and tested in the Ling Temco Vought High Speed Wind Tunnel during the first half of the 1960s. The AM 115 inlet program documents the evolution of the streamline tracing technique from an empirical process through two-dimensional characteristic solutions to the method described in Chapter 2. The series ended with the G model, which, together with the E and F models, is discussed in Ref. 13. All three models had the same design Mach number ($M_D = 2.7$), the same capture area, starting doors in each side approximately midway between leading edge and throat station, and leading-edge wedge angles that varied from 3 deg on the top (V slot) to 6 deg on the sides. Furthermore, all three inlets had conical subsonic diffusers designed in accordance with the precepts of Chapter 3 and an annular bleed slot in the throat between the subsonic and supersonic diffusers. Additional bleed was extracted through flush holes located in the forward region of the supersonic diffuser where the leading-edge shocks impacted on the opposite walls. Tests were conducted in a narrow Mach number band at and above the design value. Doors were open during starting and closed for normal operation. Bleed flow rates through the forward holes were approximately 1% and through the throat slot, as measured by metering nozzles, about 1.5%. Kinetic energy efficiencies for the main inlet flow were of the order of 0.92–0.94, which, as shown in Fig. 6.24 in the Mach number range of 2.6–3.2, encompasses the middle dashed curve (i.e., the conservatively constrained design). Thus, members of the three-dimensional inlet family can be designed in accordance with relatively uncomplicated procedures (Chapters 2 and 3) and can achieve total pressure recovery levels that are competitive with the performance of two-dimensional and axisymmetric inlets for application to practical missile configurations.

The next step in the determination of total performance capability for any inlet is the establishment of a numerical link between total pressure recovery and external drag. The kinetic energy efficiency concept explored in the several preceding paragraphs provides a convenient basis for developing an equation. Derivation of the equation proceeds from a statement of engine thrust using a modified form of Eq. (1.9):

$$F = (1 + f/a)(1 - \ell)\rho_0 A_0 V_0 V_6 + P_6 A_6 - (\rho_0 A_0 V_0^2 + P_0 A_0) \tag{6.35}$$

where ℓ is leakage and usually represents any bleed flow extracted from the inlet.

This is an expression for net jet thrust (no drag term) and may be nondimensionalized by means of a thrust coefficient as defined in Chapter 1:

$$C_{F_{NJ}} = \frac{(1 + f/a)(1 - \ell)\rho_0 A_0 V_0 V_6}{\frac{1}{2}\rho_0 V_0^2 A_C} + \frac{P_6 A_6}{\frac{1}{2}\rho_0 V_0^2 A_C} - \frac{P_0 A_6}{\frac{1}{2}\rho_0 V_0^2 A_C} - \frac{\rho_0 A_0 V_0^2}{\frac{1}{2}\rho_0 V_0^2 A_C} \tag{6.36}$$

where the thrust has been referenced to ambient pressure P_0, and the reference area for the thrust coefficient is A_C. With some manipulation, the expression finally becomes

$$C_{F_{NJ}} = \underbrace{2\frac{A_0}{A_C}\left[\left(1 + \frac{f}{a}\right)(1 - \ell)\frac{V_6}{V_0} - 1\right]}_{\text{momentum term}} + \underbrace{\frac{A_6}{A_C}\frac{P_0}{q_0}\left[\frac{P_6}{P_0} - 1\right]}_{\text{pressure term}} \tag{6.37}$$

If the exit flow is expanded to ambient pressure, $P_6 = P_0$, the pressure term above becomes zero. Also, by definition, $V_6 = C_v V_{6_{\text{ideal}}}$ and $V_0 = a_0 M_0$ [Eq. (1.11)], where C_v is the velocity coefficient and equals the nozzle efficiency n_N, as defined in Eq. (1.16) at the nozzle design pressure ratio.

From the energy equation [Eq. (1.4)],

$$V_{6_{\text{ideal}}} = \sqrt{2gJ(H_6 - h_6)} = \sqrt{2gJH_6\,(1 - h_6/H_6)}$$

but

$$\frac{h_6}{H_6} = \frac{C_p T_6}{C_p Tt_6} = \left(\frac{P_6}{Pt_6}\right)^{\gamma_6 - 1/\gamma_6} = \left(\frac{P_0}{Pt_6}\right)^{\gamma_6 - 1/\gamma_6}$$

Also, from the energy equation,

$$V_0 = \sqrt{2gJH_0\left[1 - \frac{P_0}{Pt_0}^{\gamma - 1/\gamma}\right]}$$

Substituting the preceding conditions, definitions, and equalities into Eq. (6.37) gives

$$C_{F_{NJ}} = 2\frac{A_0}{A_C}\left\{\left(1 + \frac{f}{a}\right)\left(1 - \ell\right)C_v\left\{\frac{H_6}{H_0}\frac{\left[1 - \left(\frac{P_0}{Pt_6}\right)^{\gamma_6 - 1/\gamma_6}\right]}{\left[1 - \left(\frac{P_0}{Pt_0}\right)^{\gamma - 1/\gamma}\right]}\right\}^{1/2} - 1\right\} \tag{6.38}$$

The inlet kinetic energy efficiency previously defined may be expressed as

$$n_{kE_I} = \frac{V_{2'}^2}{V_0^2} = \frac{2gJH_2\left(1 - \dfrac{h_2}{H_2}\right)}{2gJH_0\left(1 - \dfrac{h_0}{H_0}\right)} = \frac{H_2\left[1 - \left(\dfrac{P_0}{Pt_2}\right)^{\gamma - 1/\gamma}\right]}{H_0\left[1 - \left(\dfrac{P_0}{Pt_0}\right)^{\gamma - 1/\gamma}\right]} \tag{6.39}$$

where $H_2 = H_0$. Similarly, an n_{kE} for the combustor in terms of pressure change and an n_{kE} for the combustor in terms of heat addition, or γ change, can be defined so that

$$n_{kE_I}\, n_{kE_c} = \frac{\left[1 - \left(\dfrac{P_0}{Pt_6}\right)^{\gamma_6 - 1/\gamma_6}\right]}{\left[1 - \left(\dfrac{P_0}{Pt_0}\right)^{\gamma - 1/\gamma}\right]} \tag{6.40}$$

and

$$C_{F_{NJ}} = 2\,\frac{A_0}{A_C}\left\{\left(1 + \frac{f}{a}\right)(1 - \ell)C_v\left[\frac{H_6}{H_0}\, n_{kE_I} \times n_{kE_c}\right]^{1/2} - 1\right\} \tag{6.41}$$

but, by definition,

$$C_v^2 = n_{kE\ \text{nozzle}} \tag{6.42}$$

and finally,

$$C_{F_{NJ}} = 2\,\frac{A_0}{A_C}\left\{\left(1 + \frac{f}{a}\right)(1 - \ell)\left[\frac{H_6}{H_0}\, n_{kE_I} \times n_{kE_c} \times n_{kE_N}\right]^{1/2} - 1\right\} \tag{6.43}$$

The net thrust coefficient is defined as the net jet thrust coefficient minus the external drag coefficient. Mathematically,

$$C_{F_N} = C_{F_{NJ}} - C_{D_E} \tag{6.44}$$

Substituting Eq. (6.43) into Eq. (6.44),

$$C_{F_N} = 2\,\frac{A_0}{A_C}\left\{\left(1 + \frac{f}{a}\right)(1 - \ell)\left[\frac{H_6}{H_0}\, n_{kE_I} \times n_{kE_c} \times n_{kE_N}\right]^{1/2} - 1\right\} - C_{D_E} \tag{6.45}$$

Defining an effective n_{kE} where C_{D_E} is incorporated with the inlet n_{kE},

$$C_{F_N} = 2\frac{A_0}{A_C}\left\{\left(1 + \frac{f}{a}\right)(1 - \ell)\left[\frac{H_6}{H_0} n_{kE_{\text{eff}}} n_{kE_c} n_{kE_N}\right]^{1/2} - 1\right\} \tag{6.46}$$

Substituting Eqs. (6.46) and (6.43) into Eq. (6.44) and performing the appropriate steps yields

$$(n_{kE_{\text{eff}}})^{1/2} = (n_{kE_I})^{1/2} - \frac{C_{D_E}}{2\frac{A_0}{A_C}\left(1 + \frac{f}{a}\right)\left(1 - \ell\right)\left(\frac{H_6}{H_0} n_{kE_{CN}}\right)^{1/2}} \tag{6.47}$$

Finally,

$$n_{kE_{\text{eff}}} = n_{kE_I} - \frac{C_{D_E}(n_{kE_I})^{1/2}}{\frac{A_0}{A_C}\left(1 + \frac{f}{a}\right)\left(1 - \ell\right)\left(\frac{H_6}{H_0} n_{kE_{CN}}\right)^{1/2}} + \frac{C_{D_E}^{\,2}}{4\left(\frac{A_0}{A_C}\right)^2\left(1 + \frac{f}{a}\right)^2\left(1 - \ell\right)^2 \frac{H_6}{H_0} n_{kE_{CN}}} \tag{6.48}$$

For cold flow,

$$f/a = 0, \qquad H_6 = H_0, \qquad n_{kE_{CN}} = 1.0$$

$$n_{kE_{\text{eff}}} = n_{kE_I} - \frac{C_{D_E}(n_{kE_I})^{1/2}}{\frac{A_0}{A_C}(1 - \ell)} + \frac{C_{D_E}^{\,2}}{4\left(\frac{A_0}{A_C}\right)^2(1 - \ell)^2} \tag{6.49}$$

where n_{kE_I} is determined from Eq. (6.34), and C_{D_E} can be all or any part of the engine drag coefficient as measured or determined by the procedures presented in Section 6.3. For consistency the drag coefficient must be referenced to the inlet cowl area A_C.

Many pressure recovery/drag tradeoff situations are possible, but the three most commonly encountered in fixed-geometry inlet design will serve to illustrate the use of Eq. (6.49) to find a balance or best compromise. The three situations involve the following:

1) Increasing critical total pressure recovery by increasing the supersonic diffuser compression surface angle. For a flow-aligned cowl lip, the external cowl drag also increases.

2) Increasing critical total pressure recovery by extracting increasing

amounts of flow from the inlet. This bleed flow contributes to external drag.

3) Maintaining a pressure recovery level while decreasing cowl drag by flattening the cowl. Bleed is usually required to control the internal cowl lip shock/boundary-layer interactions, and bleed drag can offset any reduction in cowl drag.

Situation 1 is best exemplified by a series of three inlets designed at Mach 3.0, which is the midpoint in the freestream Mach number range of interest. Salient geometrical features of the three inlets are shown in Fig. 6.25, where the configurations are identified as the single-surface 15-deg wedge, the conservatively constrained semi-isentropic surface, and the turning limit semi-isentropic surface, which were originally introduced without specific design features in Chapter 2. All configurations are fixed-geometry, all-external compression with an internal cowl lip angle aligned with the local flow and external angle 5 deg greater than the internal angle. No bleed flow is extracted from any inlet.

Performance characteristics of the three inlets at the design Mach number only are summarized in Table 6.13. The critical total pressure recovery and corresponding kinetic energy efficiency were taken from Fig. 6.24, and the cowl wave drag coefficients were determined by the procedures presented in Section 6.3 using inlet layouts of the same scale previously used for the Mach 3.0 example inlet. The normal shock inlet is included as a reference.

The solution to Eq. (6.49) for each of the four inlets shown in the table is seen in Table 6.14, and the results are shown graphically in Fig. 6.26, where effective kinetic energy efficiency is plotted against compression

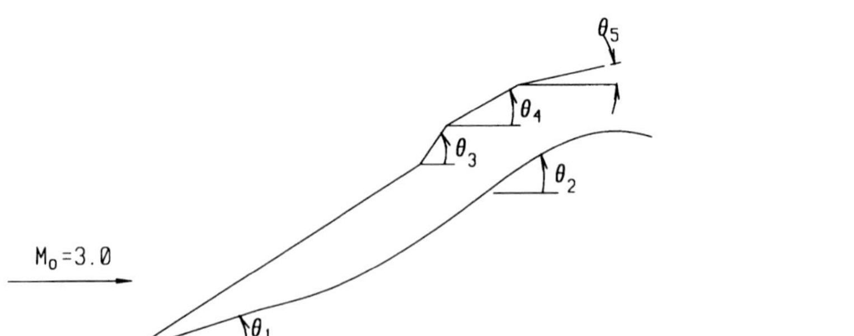

CONFIGURATION	COMPRESSION SURFACE INITIAL ∡ θ_1	FINAL θ_2	EXTERNAL COWL LIP θ_3	TAPER #1 θ_4	TAPER #2 θ_5
SINGLE WEDGE	15	15	20	8	0
SEMI-ISENTROPIC (CONSERVATIVE)	5	23.5	28.5	15	6
SEMI-ISENTROPIC (TURNING LIMIT)	5	34.5	39.5	20	12

Fig. 6.25 Salient geometrical features of three inlets designed at Mach 3.0.

Table 6.13 Summary of performance characteristics for a series of fixed-geometry all-external compression inlets designed at Mach 3.0

M_D = 3.0 inlet	Capture area ratio A_0/A_C	Critical total pressure recovery Pt_2/Pt_0	Kinetic energy efficiency n_{kE_I}	Cowl drag coefficient $C_{DE} = C_P \times dA/A_C$	Leakage (bleed) ℓ
Normal shock	1.0	0.328	0.792	0	0
15-deg Wedge	1.0	0.486	0.873	0.031	0
Conservative turning	1.0	0.650	0.927	0.084	0
Limit turn	1.0	0.797	0.963	0.452[a]	0

[a]Shock detached from cowl lip

surface final angle. The curve does peak at the conservatively constrained configuration as stipulated previously, although the 15-deg wedge inlet is certainly competitive. The inlet designed at the compressive turning limit is far worse than the normal shock inlet, primarily because the shock is detached from the cowl lip, and the cowl lip drag becomes excessively high.

Situation 2 is best exemplified by the conservatively constrained Mach 3.0 example inlet equipped with flush holes in the side walls (the first case in the bleed drag development of Section 6.3) and equipped with a ram slot (the second case in Section 6.3). The bleed geometries are shown in Figs. 6.20 and 6.21, and bleed performance is given in Fig. 6.22. It is postulated that inlet critical total pressure recovery will increase with

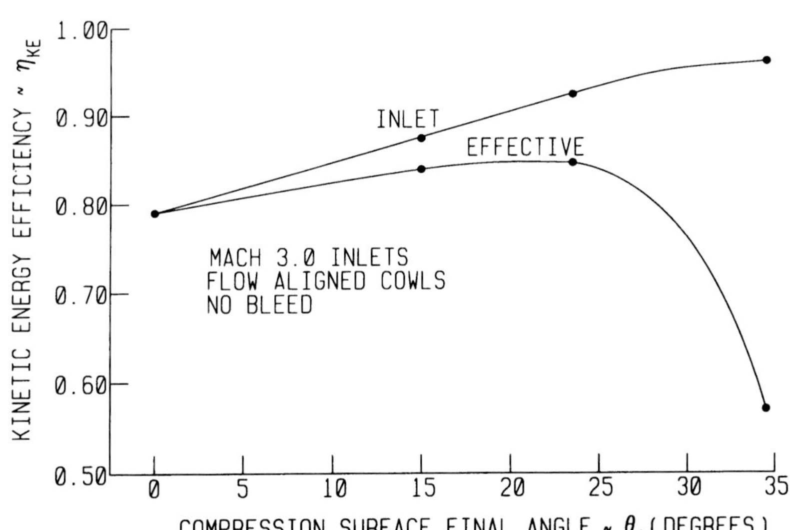

Fig. 6.26 Variation of kinetic energy efficiency with compression surface angle.

Table 6.14 Calculation of effective kinetic energy efficiency for a series of fixed-geometry all-external compression inlets designed at Mach 3.0

	n_{kE_I}	$-$	C_{D_E}		$\sqrt{n_{kE_I}}$	$+$	$C_D^2/4$	$=$	$n_{kE_{\text{eff}}}$
NS	0.792	$-$	0		0	$+$	0		0.792
Wedge	0.873		0.031	×	0.934		0.0002		0.844
Conservative	0.927		0.084	×	0.963		0.0018		0.848
Limit	0.963		0.452	×	0.981		0.0511		0.571

increasing bleed flow rate as shown in Fig. 6.27, where the trend of the curve matches the general trend recorded for the several inlets in Fig. 4.14. Performance characteristics for the two bleed cases and the no bleed reference are summarized below in Table 6.15, and the solution to Eq. (6.49) can be seen in Table 6.16.

The results are plotted in Fig. 6.28, where effective n_{kE} is plotted against bleed flow rate. For the postulated bleed flow rates, bleed recovery levels, and inlet critical total pressure recovery attainable as a result of bleed, the curve has a minimum point at the flush hole case. The drag associated with the zero momentum recovery of the flush holes in the side walls more than offsets the increased critical total pressure recovery attributed to the bleed, and the effective result is worse than the zero bleed case. The ram slot bleed case shows the best trade, but the postulated bleed recovery may be unattainable in practice. If the bleed recovery is of the order of that determined from measurements in the shrouded two-dimensional inlet tests, then the effective n_{kE} of the ram slot case would be about the same as that in the zero bleed case. The level of bleed arrangement recovery actually attainable in practice should be the deciding factor in the selection of bleed vs zero bleed, if all other conditions remain unchanged. Determination of the actual bleed recovery level of any configuration generally requires testing.

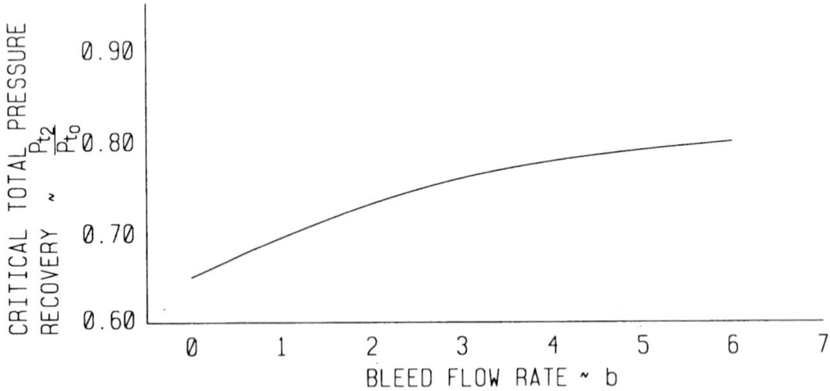

Fig. 6.27 Estimated variation of critical total pressure recovery with bleed flow rate for a Mach 3.0 design inlet.

Table 6.15 Summary of performance characteristics for the conservatively constrained Mach 3.0 design example inlet with three different bleed geometries

$M_D = 3.0$ inlet	$\frac{A_0}{A_C}$	$\left(\frac{Pt_2}{Pt_0}\right)_{crit}$	b	C_{D_B}	C_{D_C}	$C_{D_{total}}$	n_{kE_I}
No bleed	1.0	0.650	0	0	0.084	0.084	0.927
Flush holes	1.0	0.735	0.023	0.046	↓	0.130	0.949
Ram slot	1.0	0.800	0.061	0.006	↓	0.090	0.963
		Fig. 6.27	Fig. 6.22		Fig. 6.17		

Table 6.16 Calculation of effective kinetic energy efficiency for the conservatively constrained Mach 3.0 design example inlet with three different bleed geometries

	n_{kE_I}	$-$	$\frac{C_{D_E}}{(1-b)}\sqrt{n_{kE_I}}$	$+$	$\frac{C_{D_E}^{\,2}}{4(1-b)^2}$	$=$	$n_{kE_{eff}}$
No bleed	0.927	$-$	0.084×0.963	$+$	0.0018		0.848
Flush holes	0.949		$\frac{0.130 \times 0.974}{0.977}$		0.0044		0.824
Ram slot	0.963		$\frac{0.090 \times 0.981}{0.939}$		0.0023		0.871

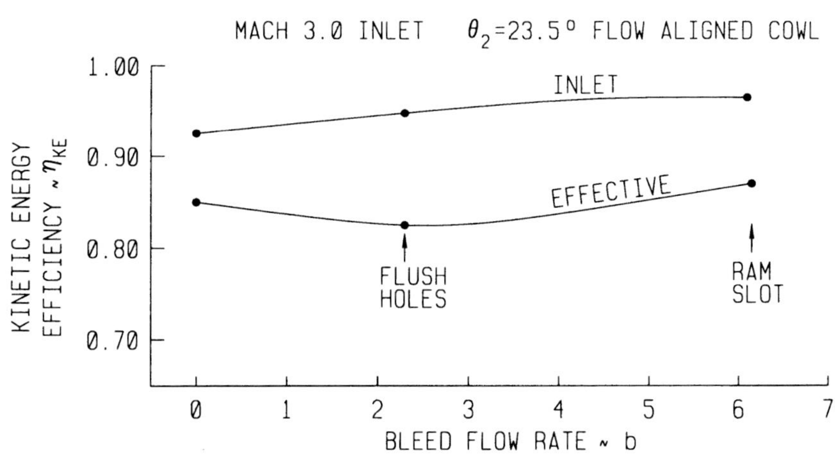

Fig. 6.28 Variation of kinetic energy efficiency with bleed flow rate.

Situation 3 is best exemplified by two versions of the shrouded two-dimensional inlet, which provided the example for the calculation of angle-of-attack performance. The first version, which has been tested in a wind tunnel, as shown in the sketch in Fig. 6.29a without the shroud, features a flattened cowl and a flush slot bleed entrance in the throat section. The second version, shown in Fig. 6.29b, is postulated to have a flow aligned cowl and no bleed. Both inlets are designed at Mach 2.5, and the salient geometrical features are given in the figure.

The performance characteristics of the two versions are summarized in Table 6.17 over a range of freestream Mach numbers. Both are assumed to have the critical total pressure recovery values shown in Fig. 6.8 at 0-deg α, and the capture area ratio values shown in Fig. 6.12. Cowl wave drag coefficients were determined by the procedures of Section 6.3 using large-scale inlet layouts to determine angles and projected frontal area. The flattened cowl version was assumed to require twice the bleed flow rates shown in Fig. 6.22 in order to achieve the desired critical total pressure recovery values, but the bleed recovery was assumed to be twice the values determined from measurements. The effective kinetic energy efficiencies of the two versions are listed in Table 6.18.

The results are given in Fig. 6.30, where effective n_{kE} is plotted against freestream Mach number for the two versions. Obviously, the flattened cowl version with bleed is better than the no-bleed aligned cowl version at all Mach numbers. The margin of superiority shown for version 1 would be completely eroded if the axial momentum of the postulated bleed flow were zero (i.e., $C_{D_B} = 2b$). Thus, any reasonably crafted bleed arrangement in consonance with the flattened cowl should have a better drag/pressure recovery tradeoff than the flow-aligned cowl version with the same critical total pressure recovery capability.

In summary, there is a procedure to weigh the opposing effects of total pressure recovery and external drag and strike a balance that produces the most efficient conversion of available energy to useful energy. The example inlets, used herein to illustrate the application of the procedure in three common situations, represent realistic geometrical configurations, but the performance postulated for some of the cases and versions may be unattainable in practice. In particular, neither the total pressure recovery of a

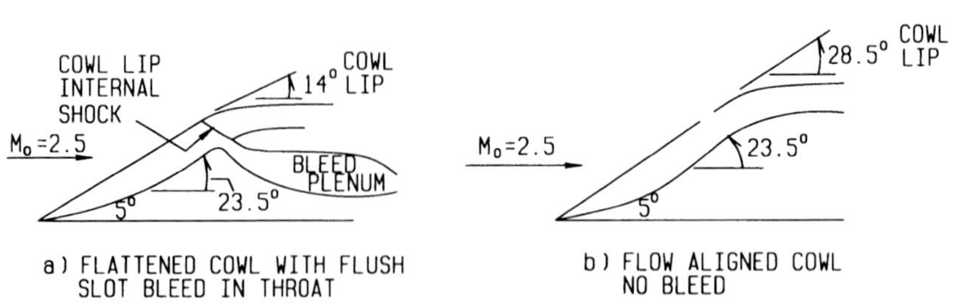

Fig. 6.29 Salient geometrical features of a two-dimensional inlet designed at Mach 2.5 showing a flattened cowl version that was tested in a wind tunnel and a postulated flow-aligned cowl version.

Table 6.17 Summary of performance characteristics for the shrouded two-dimensional Mach 2.5 design inlet with two different cowl/bleed versions

	Both versions		Flattened cowl 1, Flush slot bleed					Flow-aligned cowl 2, No bleed		
M_0	$\frac{A_0}{A_C}$	$\left(\frac{Pt_2}{Pt_0}\right)_{crit}$	n_{kE_I}	b	C_{D_B}	C_{D_C}	$C_{D_{total}}$	b	C_{D_B}	C_{D_C}
1.80	0.710	0.900	0.953	0.100	0.027	0.013	0.040	0	0	0.212
2.16	0.829	0.872	0.957	0.122	0.033	0.019	0.052	↓	↓	0.266
2.50	1.000	0.793	0.945	0.092	0.042	0.017	0.059	↓	↓	0.171
2.95	↓	0.500	0.874	0.054	0.037	0.015	0.052	↓	↓	0.145
3.50	↓	0.282	0.822	0.056	0.034	0.014	0.048	↓	↓	0.132
3.95	↓	0.186	0.802	0.058	0.033	0.014	0.047	↓	↓	0.125

Table 6.18 Calculation of effective kinetic energy efficiency for the shrouded two-dimensional Mach 2.5 design inlet with two different cowl/bleed versions

Flattened cowl version 1 Flush slot bleed

$$n_{kE_I} - \frac{C_{D_E}}{\frac{A_0}{A_C}(1-b)}\sqrt{n_{kE_I}} + \frac{C_{D_E}^{2}}{4\left(\frac{A_0}{A_C}\right)^{2}(1-b)^{2}} = n_{kE_{\mathrm{eff}}}$$

M_0									
1.80	0.953	−	0.0395 ×	0.976/(0.710 ×	0.9)	+	0.0010	=	0.893
2.16	0.957		0.0523	0.978/0.829	0.878		0.0013		0.888
2.50	0.945		0.059	0.972/1.00	0.908		0.0010		0.883
2.95	0.874		0.052	0.935/	0.946		0.0008		0.824
3.50	0.822		0.048	0.907/	0.944		0.0006		0.777
3.95	0.802		0.047	0.896/ ↓	0.942		0.0006		0.759

Aligned cowl version 2 No bleed

1.80	0.953	−	0.212 ×	0.976/(0.710 ×	1.00)	+	0.0223	=	0.683
2.16	0.957		0.266	0.978/0.829			0.0258		0.669
2.50	0.945		0.171	0.972/1.00			0.0073		0.786
2.95	0.874		0.145	0.935/			0.0053		0.743
3.50	0.822		0.132	0.907/			0.0044		0.706
3.95	0.802		0.125	0.896/ ↓	↓		0.0039		0.694

given bleed arrangement nor the slope of the inlet critical total pressure recovery vs bleed flow rate curve is analytically predictable with credible accuracy. Thus, the results obtained herein, which generally favor the extraction of significant (6–12%) amounts of bleed flow, should not be arbitrarily extrapolated to any inlet in any situation, despite geometrical

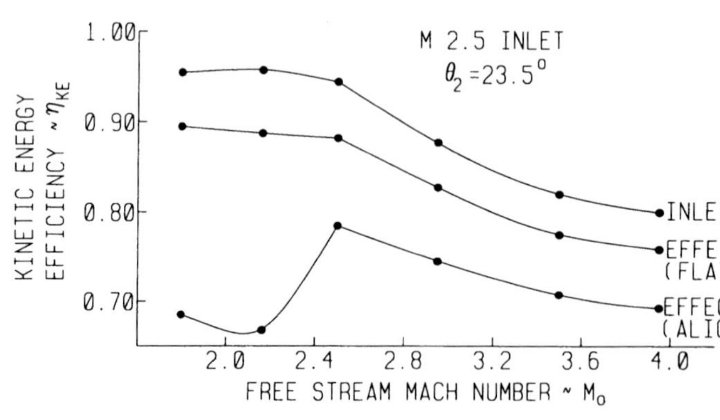

Fig. 6.30 Variation of kinetic energy efficiency with freestream Mach number for two versions of a Mach 2.5 inlet.

similarities. For most accurate results where bleed is involved, it is prudent to conduct exploratory tests, where such factors as bleed flow rate, bleed arrangement geometry, and bleed recovery are systematically varied while inlet capture area ratio, critical total pressure recovery, and exit profiles are being carefully measured. If possible, a force model of the inlet configuration with provisions to open or close the bleed passage should be tested to determine the drag contribution of the bleed flow and the overall drag coefficient of the inlet.

7
INSTALLATION FACTORS

7.1 Constraints and Requirements

In the Section 1.6 inlets were classified into nine major categories, with a variety of subdivisions within each category. Each of the category subdivisions constitutes an independent design variable, and the myriad of possible combinations of these variables presents an imposing array of options available to the inlet designer. However, availability does not translate into unrestricted freedom to choose a "best" inlet configuration for any given application. The inlet selection process will always be tempered by the requirements associated with the other disciplines involved in the development of a total system, and the number of realistic inlet options will be limited by the mission profile. Throughout the text, a number of constraints have been imposed on the inlet design variables. The initial constraint on operational Mach number range (2.0–4.0), and subsequent restrictions on family (two-dimensional or axisymmetric), geometry (fixed only), supersonic diffuser form (primarily all external), and supersonic compression complexity (emphasis on single isentropic surface) combine to yield an inlet configuration that is usually observed in a typical installation on an unmanned missile sustained by a ramjet engine and boosted to ramjet takeover speed by a rocket engine. Such missiles may be divided into two general classes: surface-launched and air-launched.

Surface-launched missiles are typically "large" with few constraints on weight, volume, or inlet placement on the vehicle. The land-launched BOMARC, which was operational with the USAF, and the ship-launched TALOS, which was operational with the USN, are representative of this genre. Both missiles are boosted by a rocket engine that imparts a velocity increment of about 2500 ft/s. Both missiles employ the parallel rocket/ramjet concept wherein the boost and sustain engines are separate entities independently designed and, in the BOMARC configuration, can be operated simultaneously. The relative size of the two surface-launched missiles and some air-launched missiles to be described subsequently is shown in Fig. 7.1. As a reference point, the airplane-like BOMARC is over 45 ft long, weighs 15,000 lbs at launch, and is powered by two 28-in.-diam ramjet engines mounted on pylons so that the inlets operate in the freestream unaffected by any vehicle-generated flowfield. BOMARC and TALOS

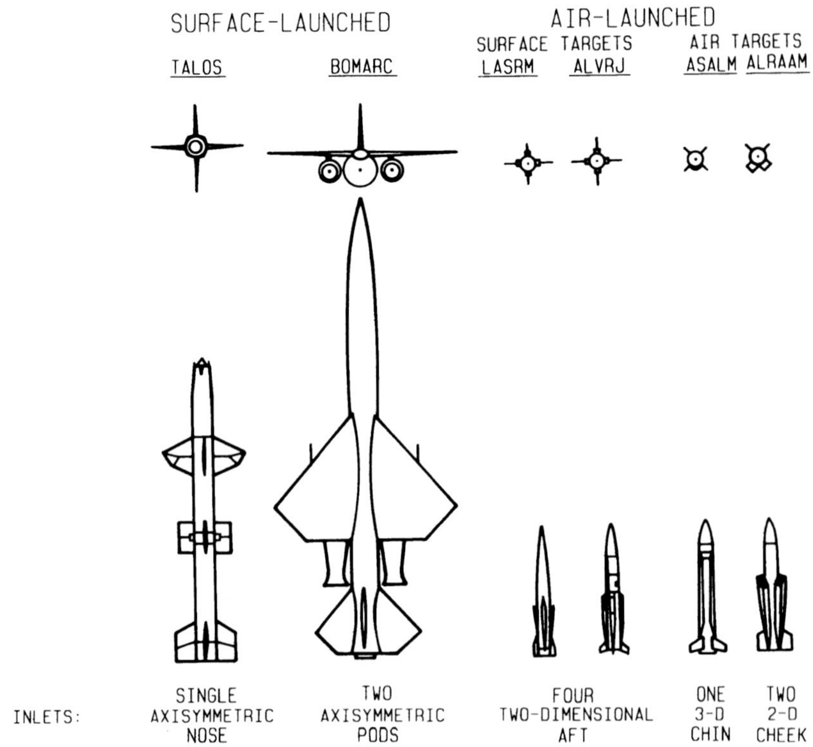

Fig. 7.1 Representative missiles powered by boost rocket/sustain ramjet engines.

have axisymmetric inlets in line with the combustors/nozzles to form flow-through ramjet engines. The engines are large (i.e., low thrust per unit frontal area), because the allowable heat release was limited by the unavailability of efficient structural cooling techniques with lightweight components.

Ramjet-powered air-launched missiles became feasible with the introduction of the integral (series) rocket/ramjet concept, the discovery of an efficient insulation material, and the determination that offset inlets closely integrated with elements of the vehicle could provide acceptable performance. These technological advances reduced the overall missile weight and volume requirements for a given payload size, increased the thrust per unit frontal area capability, and permitted inlet placement tailored to the carry aircraft launcher constraints. The integral (series) rocket/ramjet is so named because rocket booster and ramjet sustainer share the same combustion chamber, and the rocket and ramjet must be operated in series with a transition sequence between propulsive modes. Port covers are required to seal the combustion chamber from the inlet during rocket operation. Generally, for efficient booster performance a separate rocket nozzle, held inside the ramjet nozzle, is used. At rocket grain burnout, the rocket nozzle and port covers are expelled through the ramjet nozzle to clear the chamber for ramjet operation. For many applications no ejecta are permitted during transition; hence, nozzleless boosters and consumable port covers are being developed. Also, in many applications the inlets are covered to prevent damage during ground handling and to reduce drag during captive carry, launch, and boost. These covers must be ejected before the end of boost in order to effect the transition from rocket to ramjet mode. The most efficient thermal protection discovered to date for a ramjet combustion chamber is the Dow-Corning silicone elastomer formulation, which has the tradename DC 93–104. This formulation gases and chars during heat transfer, but in properly applied thickness has successfully withstood gas temperatures on the order of 4000°R for more than 400 s.

Four air-launched missiles are shown in Fig. 7.1 to the right of the two surface-launched missiles previously described. All four have about the same overall size, which is considerably smaller than the reference BOMARC, and are an order of magnitude lower in weight than BOMARC. The missile length and diameter are constrained by the carry aircraft launcher location (i.e., external under wing, external on fuselage, or internal in a bomb bay). Each missile must fit within an imaginary box that varies in shape from rectangular under the wing to rectangular with cutouts and/or indentations on the fuselage to triangular for internal carriage on a rotary launcher such as in the B-1 or B-52 bomber. The size of each box is determined by aircraft-related features such as landing clearance with compressed landing gear struts and flat tires, adjoining missile nose to tail clearance along the fuselage, bomb bay length, diameter, and number of pie-shaped elements in a rotary launcher. The allowable missile weight is also dependent upon launcher location. For example, the allowable weight on an outboard wing launcher is usually less than that on an inboard launcher. Also, total missile weight carried by a bomber aircraft would generally be greater than that allowed on a fighter aircraft.

All four air-launched missiles shown in Fig. 7.1 represent prototypes that are powered by an integral rocket/ramjet with sufficient solid fuel rocket grain to impart a velocity increment of about 1700 ft/s. All four are designed to be launched from an aircraft flying at a Mach number on the order of 0.8. However, the two missile configurations in the middle of the figure have four inlets symmetrically located about the vehicle circumference, whereas the two missiles on the right side of the figure are asymmetric with inlets only on one side of the horizontal centerline plane. These differences are key identifiers of different missions.

The two missiles in the middle of the figure are designed for air launch against surface targets that are stationary or slow moving. The Low Altitude Short Range Missile (LASRM) was developed under Air Force cognizance in an Advanced Development Program through a series of demonstration flights. This missile is essentially a point design missile with a very narrow operating envelope: Mach number near 2.5, altitude less than one statute mile above sea level, and a range of less than 50 n.mi. with no maneuver requirement. The Advanced Low Volume Ramjet (ALVRJ) was developed under Navy cognizance in an Advanced Development Program through a series of demonstration flights. It has a much broader operating envelope than the LASRM: Mach number approaching 3.0 at altitudes 6 miles above sea level with a range up to 100 n.mi., and split level mission capability with maneuverability of up to 7-deg vehicle angle of attack with minimum thrust degradation.

The two prototype missiles on the right side of Fig. 7.1 are designed for air launch primarily against aircraft or other missiles that are capable of high-speed and high-*g* evasive maneuvers. The Advanced Strategic Air-Launched Missile (ASALM) evolved from the Multi-Purpose Missile (MPM), which had both an air-to-ground and an air-to-air mission. In the air-to-air role it would be launched from the rotary launcher of a B-52 or B-1 bomber at roughly 6 miles above sea level and would climb and accelerate to Mach numbers in excess of 3 at 16 miles above sea level at ranges approaching 200 n.mi. and with a terminal maneuver capability up to angles of attack of 25 deg. The ASALM was developed under Air Force cognizance in an Advanced Development Program through a series of demonstration flights. The Advanced Long Range Air-to-Air Missile (ALRAAM) was developed under Navy cognizance in an Exploratory Development Program through a series of ground tests in a number of facilities. In an operational mode it would be externally carried and launched from an F-14 and would have an operational envelope similar to ASALM, except that the range would be less because of more stringent weight and size constraints.

A third air-to-air missile concept (not shown in Fig. 7.1) is an Advanced Common Interceptor Missile (ACIM), which could be used by both the Navy and the Air Force on the F-14, F-15, and F-16 aircraft. This missile would have approximately half the size and weight of the ALRAAM with the same operational capability but with about half the range.

All seven missiles described in the foregoing paragraphs are boosted by a solid-propellant rocket engine and sustained by a liquid-fueled ramjet engine. Five of the missiles feature the integral engine type shown in the upper left schematic in Fig. 7.2, where the engine centerline divides the boost mode configuration from the sustain mode configuration. During the boost mode the inlets are prevented from flowing by the port covers, and the rocket grain, which has been cast in the combustion chamber over the DC 93–104 insulation, is ignited, and the products of combustion are expanded through the rocket nozzle mounted within the ramjet nozzle. At rocket grain burnout, the series of events that constitute the transition sequence is triggered. Usually, explosive bolts are fired to release the rocket nozzle insert, differential air pressure pushes the port covers through the chamber and through the ramjet nozzle, and the fuel pump begins delivery of the ramjet fuel from the tank to the injectors located in the turn section of the inlet. During the sustain mode the fuel/air mixture is ignited with a flame or spark, and the pilot flame is held in the sheltered head end dome section of the combustor. The other two missiles in Fig. 7.1 feature the parallel engine concept shown in the upper right schematic in Fig. 7.2, where the booster and sustain engines are two completely separate entities and, in the case of BOMARC, are both external to the vehicle body.

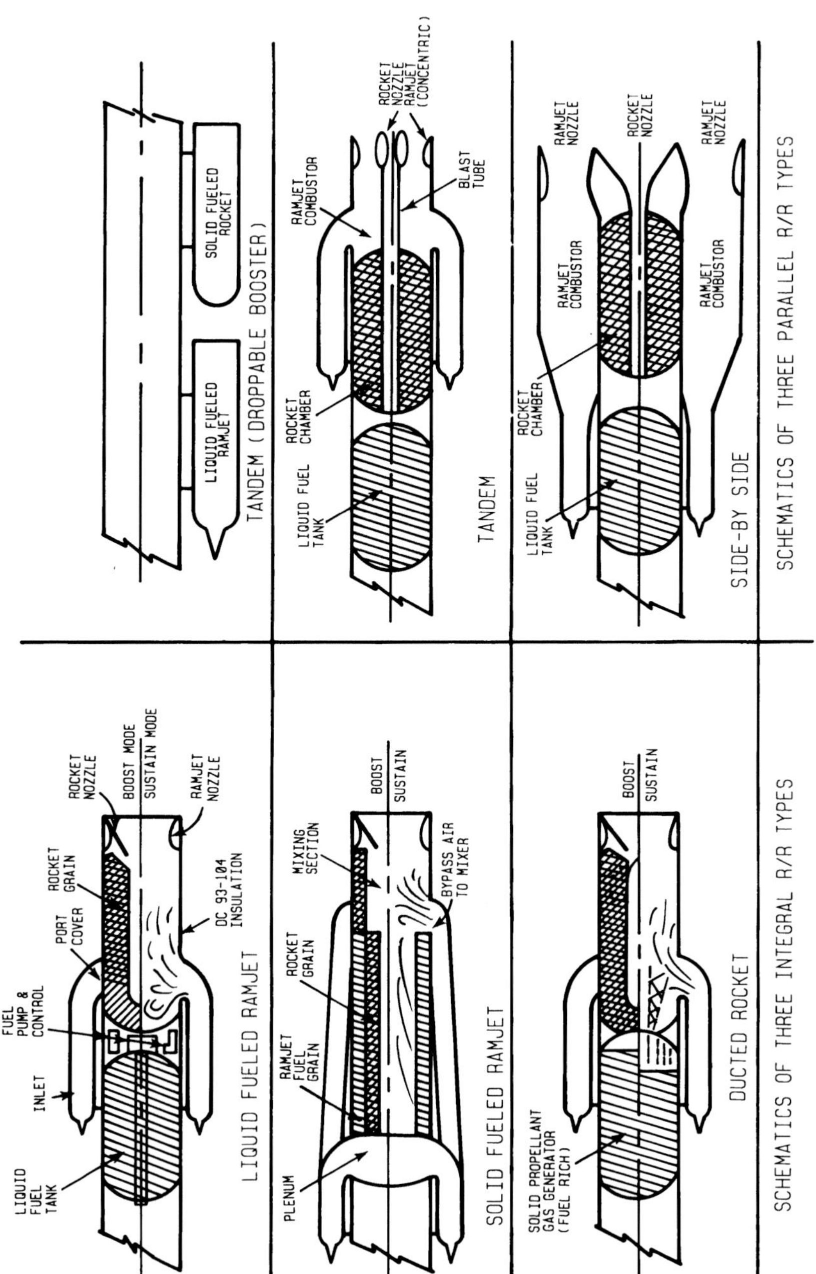

Fig. 7.2 Representative rocket/ramjet engine combinations.

Two other integral engine types (the solid-fueled ramjet and the ducted rocket) and two other parallel engine concepts (tandem and side-by-side) have been developed to various stages of maturation and are included in Fig. 7.2 to provide a more complete picture of the kinds of engine combinations that are feasible.

7.2 Inlet-Selection Variables

A typical internal packaging profile of an air-launched missile with an integral solid rocket/liquid-fueled ramjet is shown in Fig. 7.3. Also shown are four possible inlet arrangements that are viable candidates for a closely integrated configuration. These four arrangements are representative of a much larger array of candidates that include such variables as inlet family, number, placement on vehicle, and, in the case of the two-dimensional family, compression direction. The choice of a most logical candidate must be predicated on simultaneous satisfaction of launcher box constraints, mission requirements, and compatibility with other missile components. As shown in Fig. 7.3, the forward half of the missile body contains nonpropulsive components, and the aft half contains most of the propulsive components. Generally, the guidance package sits in the most forward position within the missile body with the seeker looking through the ogival nose section. Most often, the next compartment aft of the guidance section contains the payload, which would normally be a warhead. The rest of the nonpropulsive section is the control compartment, which contains the autopilot. The propulsive section houses the fuel tank, the major elements of the fuel control subsystem, and the common combustion chamber. Obviously, the air captured by the closely integrated inlet(s) must enter the combustion chamber through ports in either the head end dome or the side walls.

The skid-to-turn autopilot provides the simplest guidance concept and is consonant with a symmetrical missile that has no preferred maneuvering plane. Four inlets, equally spaced around the vehicle circumference, as shown in the top two sketches in Fig. 7.3, offer the best inlet arrangement. During vehicle angle-of-attack operation, performance of the lee side inlets can be adversely affected by separation of the crossflowing vehicle boundary layer. Inasmuch as the separation occurs initially at the aft end of the vehicle and moves forward with increasing angle of attack, inlet performance deterioration can be postponed to a higher value of angle of attack by placing the inlets in the forward position (upper left sketch) rather than the aft position (upper right sketch). The penalty for forward placement with long ducts is added weight and increased friction drag. The limit on forward position is determined by noninterference with the seeker viewing angle through the vehicle nose. Generally, the designers of conventional warheads prefer that no inlet structure shield the payload compartment so that the most forward location for an inlet to ensure maximum compatibility with other missile components would be about midbody. Flight tests with the mid-body-mounted four-inlet ALVRJ have confirmed that negligible thrust deterioration occurs at up to 7-deg vehicle angle of attack, which is about the maximum required to satisfy an air-to-surface mission. For

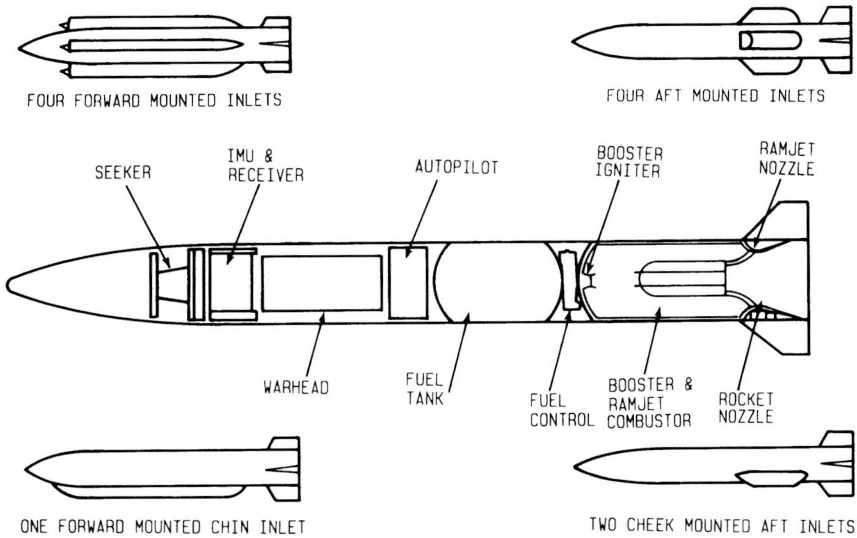

Fig. 7.3 Representative inlets on a typical vehicle showing four arrangements and internal packaging.

air-to-air missions, where high-altitude cruise and course correction maneuvers require angles of attack much higher than 7 deg with no thrust deterioration, inlets can be located only on the windward side of the body, and a bank-to-turn autopilot is required for pitch-plane-only maneuvers. Thus, air-to-air missions dictate asymmetrical configurations, as shown in the bottom two sketches in Fig. 7.3, with a more complex guidance than the symmetrical air-to-surface configurations.

The single-chin inlet configuration shown in the lower left sketch in Fig. 7.3 is tailored to fit the triangular box of the B-52 and B-1 rotary launchers. The inlet is forward-mounted to take advantage of the precompression offered by the vehicle forebody nose, and the inlet structure presents no problem for the nuclear warhead. The sketch in the bottom right of Fig. 7.3 shows a configuration with two aft inlets that are cheek-mounted to fit in the imaginary wing launcher box of the F-14. Cheek mounting provides lift, and the two-dimensional inlets use downward compression to enhance angle-of-attack operation.

Three of the four inlet arrangements in Fig. 7.3 feature two-dimensional or quasi-two-dimensional inlets, and this inlet family is more popular than the axisymmetric because of superior packaging flexibility; i.e., the height-to-width ratio of the two-dimensional inlet can be varied to adjust to different box constraints. Also, the compression direction of the two-dimensional inlet can be changed to suit the mission requirements. Variations in compression direction are shown in Fig. 7.4, where the details of three typical two-dimensional inlet installations are illustrated. Compression inward toward the vehicle body was used in the point design four-inlet

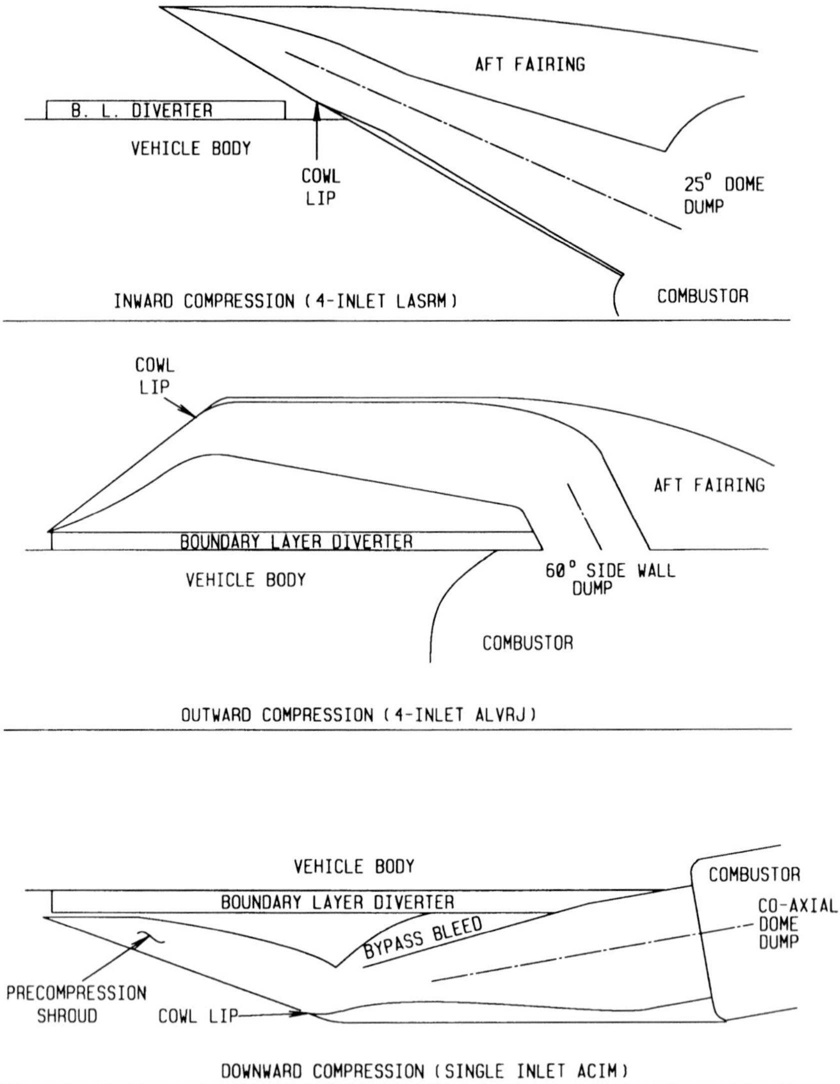

Fig. 7.4 Installation details for three two-dimensional inlets (not to scale).

LASRM configuration (top sketch in Fig. 7.4), where the lowest possible drag was the overriding requirement. The vehicle boundary-layer diverter mounted upstream from the cowl lip shielded the high-angle (high-drag) cowl surfaces from high-energy freestream flow and precluded any low-energy boundary-layer ingestion by the inlet. However, at Mach numbers below the inlet design value, impingement of the compression shocks on the diverter upper surface separated the boundary layer on that surface and usually unstarted the inlet. Subsonic diffusion is completed within the vehicle body envelope, and the compressed flow is dumped into the combustion chamber through ports in the head end dome at an angle near 25 deg.

To provide a wide operating Mach number range on either side of the design value, the four-inlet ALVRJ configuration used outward compression (middle sketch in Fig. 7.4). The inlet compression surface sits on the vehicle boundary-layer diverter so that the compression shocks interact only with freestream flow during inlet operation below design Mach number. However, the high-angle cowl lip surface is exposed directly to freestream flow; hence, the drag contribution is much higher than the inward-turning configuration. Subsonic diffusion is completed outside the vehicle body envelope, and the compressed flow is turned 60 deg and dumped through ports in the side of the combustion chamber just downstream from the head end dome.

To enhance inlet angle-of-attack capability as described in previous chapters, air-to-air missile configurations usually use downward-compressing inlets. The bottom sketch in Fig. 7.4 shows the ACIM conceptual configuration with a single inlet mounted on a vehicle boundary-layer diverter under the vehicle body aft of the ogival nose. The compressed flow is dumped through the center of the head end dome.

Generally, for any given cowl size for any mission, a single inlet would have the lowest weight and highest performance, but box constraints often dictate more than one inlet at the expense of increased weight and drag. Furthermore, at angle of attack (yaw), if more than one inlet is processing air for a common combustion chamber, the total pressure recovery of the inlet component is no better than the worst performing member. From purely a weight standpoint the axisymmetric inlet would always be preferred to an inlet of any other shape, and this inlet family was the designer's choice in the relatively unconstrained BOMARC and TALOS missiles. Also, from a weight standpoint, an inlet closely coupled with a combustor (e.g., the LASRM) would be selected over an inlet with a long transport duct (e.g., the ASALM), but short inlets can produce unacceptable flow distortions and amplify flow disturbances such as boundary-layer separations, vortices, and slip planes.

7.3 Inlet-Performance Adjustments

Performance-estimating methods presented in Chapter 6 are adequate for a complete description of the performance of in-line inlets located outside the flowfield zone of influence of all vehicle components, such as the pylon-mounted BOMARC inlets and vehicle nose-mounted TALOS inlet. For offset inlets closely integrated with a vehicle body, the inlet performance must be adjusted to incorporate external flowfield changes produced by the vehicle forebody, additional losses caused by turning and dumping the compressed flow into the combustor, and drag imposed by the vehicle boundary-layer diverter. There is also the afterbody drag caused by the fairings installed aft of most offset inlets, but this drag contribution would generally be calculated by the vehicle aerodynamicist.

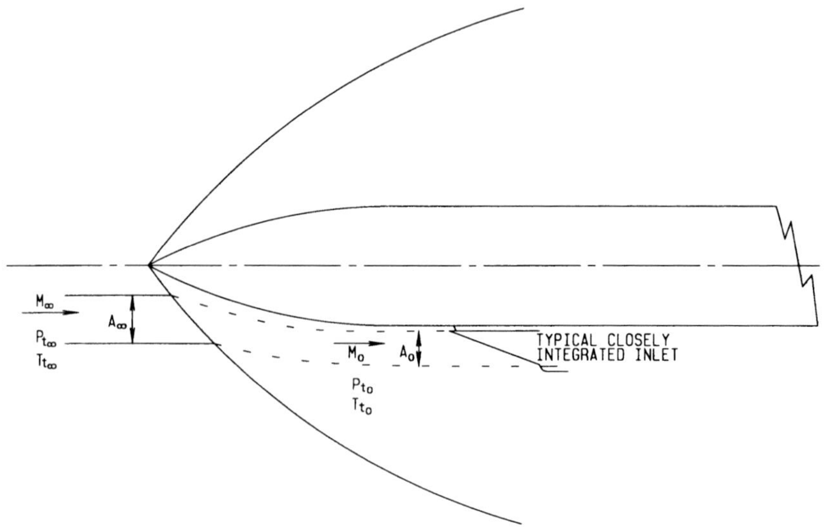

Fig. 7.5 Sketch of airflow around a typical ogive/cylinder vehicle body at 0-deg angle of attack.

A sketch of the airflow around a typical vehicle with ogival forebody operating at 0-deg angle of attack at a supersonic velocity is shown in Fig. 7.5. Freestream conditions are designated by ∞, and local flow conditions downstream from the forebody nose shock are assigned zero, which is consistent with the station number notation used in Chapter 1. The bundle of freestream tubes (with cross-sectional area A_∞) eventually captured by a closely integrated inlet must pass across the curved shock generated by the ogival forebody and sustain a change in flow properties. At the inlet face, the cross-sectional area of the captured stream tubes is A_0, which is related to A_∞ by means of the continuity equation,

$$\frac{Pt_\infty \left(\frac{P}{Pt}\mathring{m}\right)_\infty A_\infty}{\sqrt{Tt_\infty}} = \frac{Pt_0 \left(\frac{P}{Pt}\mathring{m}\right)_0 A_0}{\sqrt{Tt_0}} \tag{7.1}$$

Dividing both sides by the inlet capture area A_C, setting $Tt_\infty = Tt_0$, and rearranging to solve for the inlet capture area ratio based on freestream conditions (i.e., the installed capture area ratio),

$$\frac{A_\infty}{A_C} = \frac{A_0}{A_C}\frac{Pt_0}{Pt_\infty}\frac{\left(\frac{P}{Pt}\mathring{m}\right)_0}{\left(\frac{P}{Pt}\mathring{m}\right)_\infty} \tag{7.2}$$

where A_0/A_C is the isolated value of inlet capture area ratio as determined in Chapter 6 for any specified value of local Mach number M_0; Pt_0/Pt_∞ is the total pressure recovery across the forebody nose shock;

$$\left(\frac{P}{Pt}\mathring{m}\right)_0$$

is a function of the local Mach number M_0 at the inlet face; and

$$\left(\frac{P}{Pt}\mathring{m}\right)_\infty$$

is a function of the freestream Mach number M_∞.

The installed value of critical total pressure recovery is given by

$$\frac{Pt_2}{Pt_\infty} = \frac{Pt_2}{Pt_0}\frac{Pt_0}{Pt_\infty} \tag{7.3}$$

where Pt_2/Pt_0 is the isolated value of inlet critical total pressure recovery as determined in Chapter 6 for any specified value of local Mach number M_0.

The solutions to Eqs. (7.2) and (7.3) require knowledge of local flow properties (M_0, Pt_0) at any desired value of freestream Mach number M_∞. At 0-deg vehicle angle of attack, proven axisymmetric characteristics programs can furnish a precise definition of the complete flowfield between the body surface and forebody shock at any desired axial station or circumferential position. Inasmuch as the flow is axially symmetric about the body, the solutions to the equations are the same with the inlet rotated to any circumferential location at any given axial station. Generally, if the inlet is located downstream from the ogive/cylinder junction on the vehicle body, the average value of local Mach number M_0 is close enough to the freestream value M_∞ that the installed inlet capture area ratio can be approximated by the expression

$$\frac{A_\infty}{A_C} = \frac{A_0}{A_C}\frac{Pt_0}{Pt_\infty} \tag{7.4}$$

At vehicle angles of attack other than 0 deg, Eqs. (7.2) and (7.3) still apply, but, because the flowfield becomes three dimensional, there is a considerable variation of local flow properties both radially and circumferentially at any axial station, and the simplified characteristics methods are inadequate to describe the picture. A sketch of the airflow around the typical vehicle operating at a relatively high angle of attack, α_v, is shown in Fig. 7.6. The cross-sectional view at the inlet face station as sketched on the right side of the figure in section Z-Z illustrates the use of polar coordinates in the definition of flowfield between the body surface designated by r_0 and shock surface at r_z. Intersections of the radial surfaces designated r_1, r_2, etc., with the circumferential planes denoted ϕ_0, ϕ_{30}, etc., represent points in the flowfield with unique properties. Of particular importance to the inlet designer is the determination of the flowfield properties within the inlet capture tube boundary outlined in phantom in section *Z-Z*. Although the literature contains many references to programs that can define angle-of-attack flowfield properties, one program, developed at the Naval Surface Weapons Center (NSWC) and described in Ref. 23, has been used in practice to estimate installed inlet performance. Because the NSWC flowfield code is inviscid, the results obtained on the lee side of any body are subject to significant inaccuracy as a consequence of the adverse effects of crossflowing boundary layer. There are currently insufficient experimental data on windward side inlet configurations to determine a precise level of accuracy attainable with the flowfield code over a representative supersonic Mach number/angle-of-attack spectrum. Initial comparisons between estimates with the code and available data indicate acceptable accuracy for preliminary design calculations.

Demonstration of the use of the NSWC flowfield code in determining the installed performance of an inlet is facilitated by adopting for an example case a familiar inlet that has been previously described herein and a vehicle body whose coordinates have been input to the NSWC program. The inlet is the two-dimensional, downward-compressing shrouded configuration described in Chapter 6 with physical characteristics shown in the sketches in Figs. 6.6 and 6.7 and with isolated performance estimates shown in Figs. 6.8 and 6.12. The vehicle body is the ACIM configuration, which is a fineness ratio 2.25 Von Kármán ogive/cylinder. The inlet is bottom-mounted on the vehicle body in approximately the position shown in the sketches in Figs. 7.5 and 7.6. The steps involved in the application of the NSWC flowfield code to the selected inlet/vehicle configuration and the installed performance results are listed in Table 7.1 at vehicle angles of attack α_v of 0, 5, and 10 deg at each of five freestream Mach numbers: M_∞ = 2.0, 2.5, 3.0, 3.5, and 4.0.

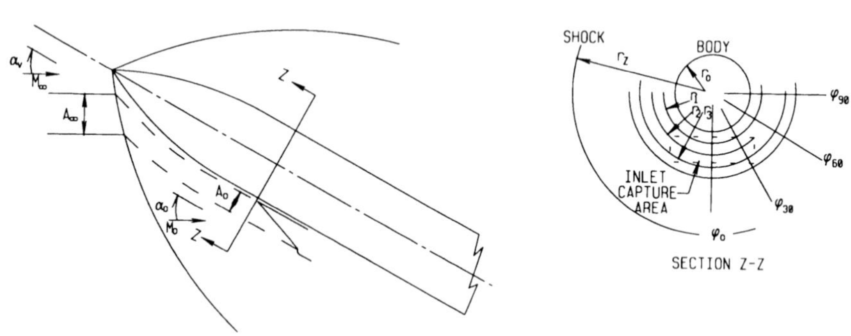

Fig. 7.6 Sketch of airflow around a typical ogive/cylinder vehicle body at a relatively high angle of attack.

Table 7.1 Calculation of critical total pressure recovery and capture area ratio for a shrouded two-dimensional inlet installed on an ACIM vehicle using the NSWC flowfield code

Freestream conditions		Inlet face average local flowfield conditions			"Corrected" isolated inlet performance		Mach number functions		Installed inlet performance	
M_∞	α_v	M_0	α_0	$\frac{Pt_0}{Pt_\infty}$	$\frac{Pt_2}{Pt_0}$	$\frac{A_0}{A_C}$	$\left(\frac{P}{Pt}\mathring{m}\right)_0$	$\left(\frac{P}{Pt}\mathring{m}\right)_\infty$	$\frac{Pt_2}{Pt_\infty}$	$\frac{A_\infty}{A_C}$
2.0	0	2.00	0.03	0.987	0.895	0.791	0.31512	0.31512	0.883	0.781
	5	2.01	3.78	0.991	0.899	0.815	0.31250	↓	0.891	0.801
	10	1.98	6.66	0.989	0.898	0.810	0.32040	↓	0.888	0.815
2.5	0	2.51	0.11	0.977	0.788	1.000	0.19981	0.20168	0.770	0.968
	5	2.50	3.57	0.977	0.826	1.070	0.20168	↓	0.807	1.045
	10	2.42	5.90	0.972	0.861	1.060	0.21724	↓	0.837	1.110
3.0	0	3.00	0.13	0.956	0.481	1.010	0.12558	0.12558	0.460	0.966
	5	2.97	2.90	0.950	0.552	1.150	0.12922	↓	0.524	1.124
	10	2.86	5.50	0.951	0.691	1.250	0.14350	↓	0.657	1.358
3.5	0	3.49	0.16	0.932	0.290	1.010	0.07905	0.07832	0.270	0.950
	5	3.43	3.17	0.931	0.352	1.170	0.08631	↓	0.328	1.200
	10	3.27	4.99	0.920	0.441	1.270	0.09719	↓	0.406	1.450
4.0	0	4.00	0.19	0.904	0.182	1.010	0.04961	0.04961	0.165	0.913
	5	3.86	3.00	0.902	0.228	1.165	0.05626	↓	0.206	1.192
	10	3.66	4.66	0.886	0.297	1.250	0.06753	↓	0.263	1.508
Arbitrarily selected		Calculated by NSWC flowfield code			↑ From Figs. 6.8	↑ 6.12	From Ref. 2		Eqs. (7.3)	(7.2)

As noted at the bottom of the table, the freestream conditions are arbitrarily selected, and the NSWC flowfield code calculates the local flowfield properties at the inlet face for each set of freestream values. The local flowfield values shown in the table are the arithmetic averages of about 20 line (r-θ) intersections within the inlet capture stream tube boundary. It is postulated that the average values are adequate for the production of acceptably accurate results and that a net fineness commensurate with 20 line intersections yields reasonably representative average values. The average values of local Mach number M_0 and local flow angle of attack α_0 are used to enter the isolated performance curves of Figs. 6.8 and 6.12 to determine "corrected" values of critical total pressure recovery Pt_2/Pt_0 and capture area ratio A_0/A_C, respectively. The installed values of critical total pressure recovery Pt_2/Pt_∞ are determined from the solution of Eq. (7.3) using the corrected isolated total pressure recovery values and the average values of total pressure recovery across the forebody shock Pt_0/Pt_∞. Installed values of capture area ratio A_∞/A_C are determined from Eq. (7.2) using the corrected isolated capture area ratios, the forebody shock recovery averages, and the appropriate Mach number functions.

The installed performance results given in Table 7.1 are shown graphically in Figs. 7.7 and 7.8, where critical values of total pressure recovery

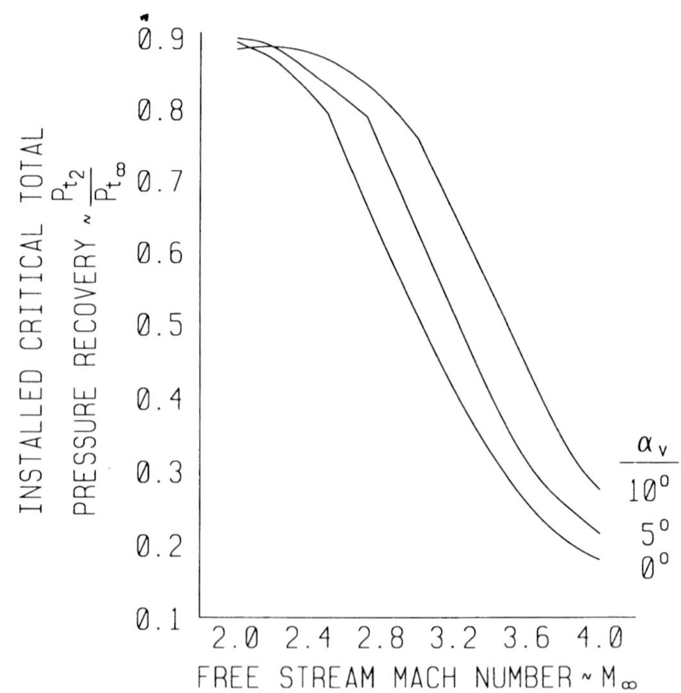

Fig. 7.7 Estimated variation of critical total pressure recovery with freestream Mach number for a shrouded two-dimensional inlet installed under an ACIM vehicle body at three angles of attack.

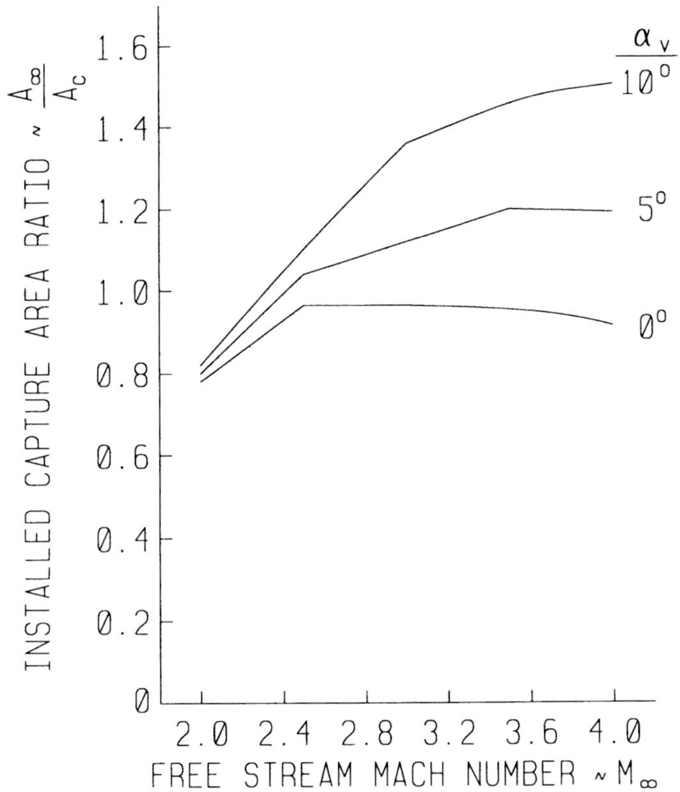

Fig. 7.8 Estimated variation of capture area ratio with free-stream Mach number for a shrouded two-dimensional inlet installed under an ACIM vehicle at three angles of attack.

Pt_2/Pt_∞ and capture area ratio A_∞/A_C, respectively, for the installed inlet are plotted against freestream Mach number M_∞ for the three values of vehicle angle of attack α_v. These estimated installed performance curves have the same trends as experimental data curves established from wind-tunnel model tests with similar two-dimensional, downward-compressing inlets mounted in about the same location under the ACIM vehicle body. Any change in the inlet selection variables or vehicle shape will produce different installed performance results, but the method for calculating the results will be as demonstrated.

Compressed air from the offset inlet is introduced into the combustor through ports in the head end dome or in the side wall. Three different port locations and flow admittance angles are shown in Fig. 7.4. For configurations such as those found in the ACIM and LASRM, where the compression process is completed at least partially within the vehicle body envelope, the flow turning is usually accomplished in the subsonic diffuser, and there is no separate, identifiable turn loss. For configurations such as

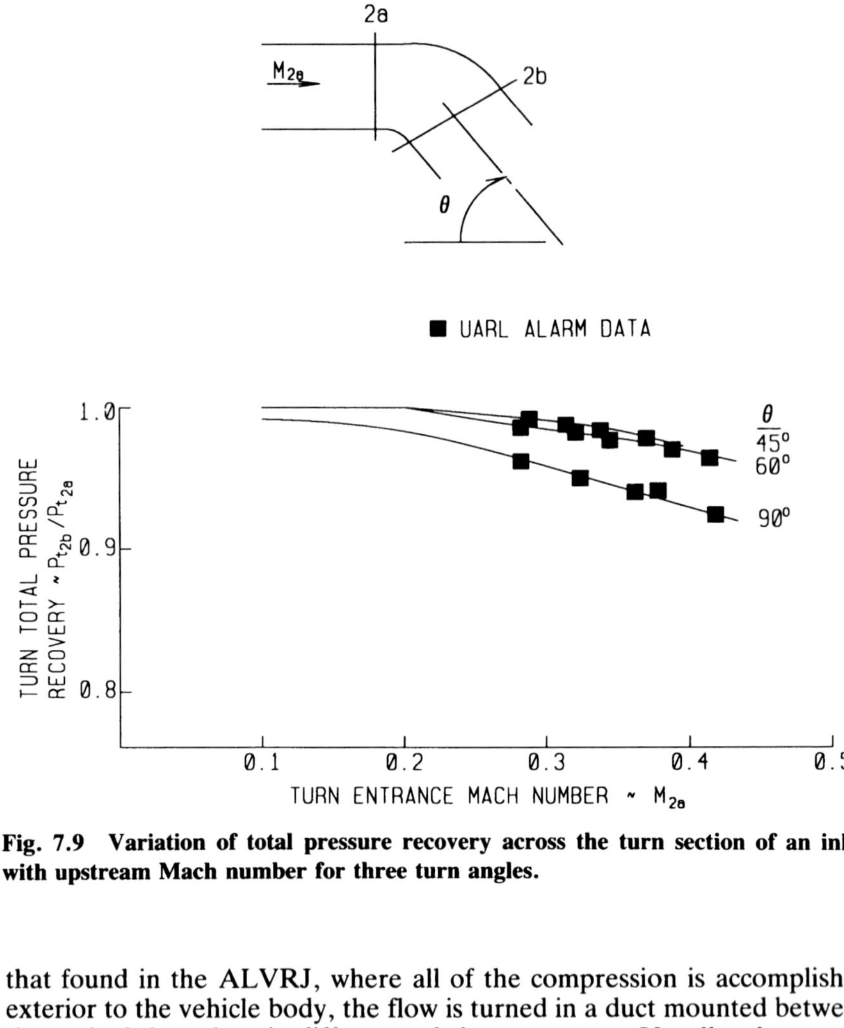

Fig. 7.9 Variation of total pressure recovery across the turn section of an inlet with upstream Mach number for three turn angles.

that found in the ALVRJ, where all of the compression is accomplished exterior to the vehicle body, the flow is turned in a duct mounted between the end of the subsonic diffuser and the entry port. Usually, the turn is accomplished in a constant area section of ducting so that boundary-layer growth will cause an acceleration of the subsonic flow to maintain favorable wall pressure gradients and preclude separation. The force required to turn the flow produces a loss in total pressure across the turn section, and the abrupt expansion of the flow from the port to the combustor with a cross-sectional area two to three times greater than the port area produces the dump loss. In the development of integral engines it has become customary to charge the combustor with any turn and dump losses; i.e., these are equated with the aerodynamic losses between stations 2 and 3 in the flow-through engine as described in Chapter 1. Although the inlet may be absolved of responsibility for turn and dump losses, it is instructive for the inlet designer to appreciate the magnitude of these losses.

In order to preserve the integrity of the station notation presented in Chapter 1, it is necessary to introduce three new substations: 2a at the beginning of the turn section, 2b at the end of the turn section, and 2c at the duct port. During the development of the inlet component for a specific engine, there is usually adequate instrumentation to define flow properties at station 2c, but no instrumentation to isolate turn losses. Under Navy sponsorship, the United Aircraft Research Laboratory conducted an applied research program to investigate some aspects of performance peculiar to offset inlets.[24] One phase of the program was devoted to tests of inlets with instrumented turn sections and hardware designed for interchangeability of sections with different turn angles. The results are presented in Fig. 7.9, where the total pressure recovery across the turn section, illustrated in the sketch, is plotted against the turn entrance Mach number for three turn angles. At any given turn angle, the total pressure recovery decreases with increasing Mach number. At any given Mach number, the total pressure recovery is practically the same for the 45- and 60-deg turns, but decreases substantially as the flow is turned through 90 deg. Based on these results, it would seem prudent to restrict flow turning to angles 60 deg or less. The inlet developed for the ALVRJ program has a turn angle of 60 deg and a turn entrance Mach number estimated to be on the order of 0.3 at the critical operating point. According to Fig. 7.9, the ALVRJ should operate with a turn total pressure recovery of about 0.99.

The abrupt expansion of the flow from the port to the combustor (the dump) can be treated as a one-dimensional process and solved by means of the stream thrust and continuity equations presented originally in Chapter 1 under the discussion of burner aerodynamic losses. The conservation of momentum requires that the stream thrust at any combustor axial station (such as 3 in Fig. 7.10) must equal the stream thrust at the port opening

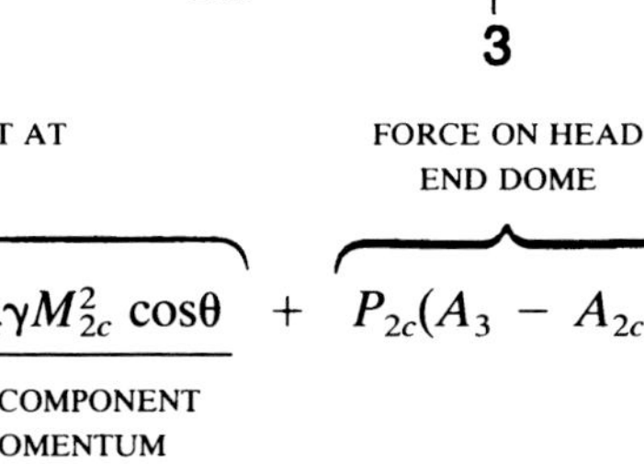

$$\overbrace{\underbrace{P_{2c}A_{2c}}_{\text{PRESSURE AREA}} + \underbrace{P_{2c}A_{2c}\gamma M_{2c}^2 \cos\theta}_{\text{AXIAL COMPONENT OF MOMENTUM}}}^{\text{STREAM THRUST AT STATION 2C}} + \overbrace{P_{2c}(A_3 - A_{2c})}^{\text{FORCE ON HEAD END DOME}} = \overbrace{P_3A_3(1 + \gamma M_3^2)}^{\text{STREAM THRUST AT STATION 3}}$$

Fig. 7.10 Sketch of a dump combustor showing an inlet port opening in the head end dome.

Table 7.2 Calculation of dump total pressure recovery for a dump flow angle of 60 deg

Arbitrarily selected		From Ref. 2	Eq. 7.7 $\theta = 60$ deg			From Ref. 2	Eq. 7.8		
$\frac{A_3}{A_{2c}}$	$M_{2c} \rightarrow$	$\mathring{m}_{2c}$	$\frac{A_3}{A_{2c}} \frac{1}{\mathring{m}_{2c}}$ +	$\frac{\frac{1}{2}\gamma M_{2c}{}^2}{\mathring{m}_{2c}}$ =	$\frac{\left(\frac{f}{P}\right)_3}{\mathring{m}_3} \rightarrow$	M_3	$\frac{A_{2c}}{A_3} \times$	$\left(\frac{P}{Pt}\mathring{m}\right)_{2c} \Big/ \left(\frac{P}{Pt}\mathring{m}\right)_3 =$	$\frac{Pt_3}{Pt_{2c}}$
10.0	0.1	0.09198	108.7193	+.0761	=108.7954	0.010	$\frac{1}{10}$	0.09134/0.00919	0.994
	0.2	0.18451	54.1976	0.1518	54.3494	0.020		×0.17944/0.01837	0.977
	0.3	0.27814	35.9531	0.2265	36.1796	0.030		0.26130/0.02755	0.949
	0.4	0.37340	26.7809	0.2999	27.0808	0.040		0.33442/0.03672	0.911
	0.5	0.47080	21.2404	0.3717	21.6121	0.050		0.39689/0.04588	0.865
	0.6	0.57085	17.5177	0.4414	17.9591	0.061		0.44754/0.05593	0.800
5.0	0.1		54.3596		54.4357	0.020	$\frac{1}{5}$	/0.01837	0.994
	0.2	S	27.0988	S	27.2506	0.040		S /0.03672	0.977
	0.3	a	17.9766	a	18.2031	0.060		a /0.05502	0.950
	0.4	m	13.3905	m	13.6904	0.080		m /0.07323	0.913
	0.5	e	10.6202	e	10.9919	0.100		e /0.09134	0.869
	0.6		8.7589		9.2003	0.120		/0.10932	0.818

2.5	0.1		27.1798		27.2559	0.040	$\frac{1}{2.5}$		/0.03672	0.995
	0.2		13.5494		13.7012	0.080			/0.07323	0.980
	0.3		8.9883		9.2148	0.120			/0.10932	0.956
	0.4		6.6953		6.9952	0.161			/0.14567	0.918
	0.5		5.3101		5.6818	0.202			/0.18115	0.876
	0.6	↓	4.3795	↓	4.8209	0.243		↓	/0.21557	0.830
2.0	0.1		21.7439		21.8200	0.050	$\frac{1}{2}$		/0.04588	0.995
	0.2		10.8395		10.9913	0.100			/0.09134	0.982
	0.3		7.1906		7.4171	0.150			/0.13599	0.961
	0.4		5.3562		5.6561	0.202			/0.18115	0.923
	0.5		4.2481		4.6198	0.255			/0.22541	0.880
	0.6	↓	3.5035	↓	3.9449	0.310		↓	/0.26905	0.832
1.0	0.1		10.8719		10.9480	0.101	1.0		/0.09224	0.990
	0.2		5.4200		5.5718	0.206			/0.18445	0.973
	0.3		3.5953		3.8218	0.323			/0.27898	0.937
	0.4		2.6781		2.9780	0.467			/0.37754	0.886
	0.5	↓	2.1240	↓	2.4957	0.706		↓	/0.48783	0.814
	0.6		—		—	—				—

(station 2c) plus any force acting on the head end dome outside the port opening. For a side dump at 90 deg, $\cos\theta = 0$, and there is no axial component of momentum. For a coaxial dump, $\cos\theta = 1.0$, and the axial component of momentum is maximum.

Combining like terms in the stream thrust expression,

$$P_{2c}A_3 + P_{2c}A_{2c}\gamma M_{2c}^2 \cos\theta = P_3A_3(1 + \gamma M_3^2) \tag{7.5}$$

and the conservation of mass requires

$$P_{2c}A_{2c}\mathring{m}_{2c} = P_3A_3\mathring{m}_3 \tag{7.6}$$

Dividing stream thrust [Eq. (7.5)] by continuity [Eq. (7.6)],

$$\frac{A_3}{A_{2c}}\frac{1}{\mathring{m}_{2c}} + \frac{\gamma M_{2c}^2 \cos\theta}{\mathring{m}_{2c}} = \frac{1 + \gamma M_3^2}{\mathring{m}_3} \equiv \frac{\left(\dfrac{f}{p}\right)_3}{\mathring{m}_3} \tag{7.7}$$

Equation (7.7) can be solved to determine the value of the Mach number at station 3 in the combustor for any number of arbitrarily selected values of A_3/A_{2c} and M_{2c} at any desired dump angle θ. As soon as M_3 is determined, the dump total pressure recovery can be calculated from the continuity equation written in the form

$$\frac{Pt_3}{Pt_{2c}} = \frac{\left(\dfrac{P}{Pt}\mathring{m}\right)_{2c} A_{2c}}{\left(\dfrac{P}{Pt}\mathring{m}\right)_3 A_3} \tag{7.8}$$

An illustrative set of calculations using Eqs. (7.7) and (7.8) for a 60-deg dump angle is listed in Table 7.2.

The total pressure recovery values listed in Table 7.2 for the 60-deg dump and values calculated for dumps at angles of 0, 30, 45, and 90 deg are plotted in Fig. 7.11. At the 0-deg dump angle (in the middle of the figure) and an area ratio of 1.0 (i.e., the flow-through engine case), there is no dump loss at any upstream dump Mach number. At every other combination of dump angle and area ratio, the total pressure recovery across the dump decreases with increasing upstream Mach number, and the rate of decrease becomes greater as the dump angle increases from 0 to 90 deg. At dump angles less than 45 deg, the recovery decreases with increasing area ratio, but at 45 deg, the area ratio unity results begin to deviate from the pattern, and at 60 and 90 deg, they show the worst performance. For these dump angles the flow must accelerate from stations 2c to 3 to satisfy the continuity requirement, and sonic flow is possible

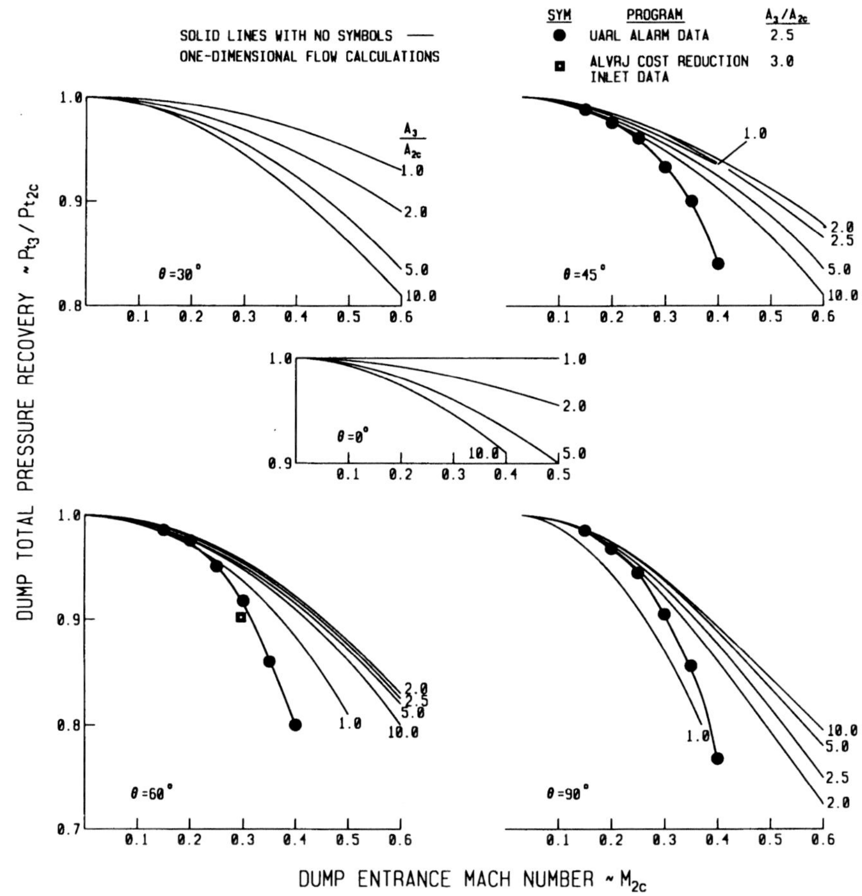

Fig. 7.11 Variation of total pressure recovery across the dump section of an inlet/combustor with upstream Mach number for five dump angles and five dump area ratios.

with high losses. At 90 deg, the total pressure recovery/area ratio relationship is completely reversed. In order to compare the analytical results determined from one-dimensional flow calculations with experimental results, measured dump total pressure ratios obtained in the UARL research program are plotted in Fig. 7.11. The measured data were acquired with a fixed dump area ratio, A_3/A_{2c}, of 2.5 at dump angles of 45, 60, and 90 deg. Generally, at upstream Mach numbers of 0.2 and below, analytical and measured results agree within the limits of experimental accuracy. However, as the upstream Mach number increases above the value of 0.2, the measured total pressure recovery decreases at a much faster rate than the analytically determined values, and the rate of decrease increases with increasing dump angle. At an upstream dump Mach number of around

0.2, a well-designed, conventional inlet would be operating at the critical condition with the terminal normal shock train contained in the throat, and the flow through the subsonic diffuser would most closely approximate the one-dimensional process. As the upstream dump Mach number increases above 0.2, indicative of supercritical operation, the flow would be subject to increasing levels of distortion with vortex sheets and boundary-layer separations adding to the total pressure losses. Measurements obtained in a cost reduction program for the ALVRJ inlet component[25] correlate quite well (at one condition) with the UARL data. The ALVRJ data point is shown at the 60-deg dump angle, upstream Mach number = 0.3 condition, and the area ratio is approximately 3.0. Note that the ALVRJ dump loss is an order of magnitude greater than the turn loss described in the discussion of Fig. 7.9. The LASRM inlet had about a 25-deg dump angle and an area ratio of 2.2. Plotted on the 30-deg dump angle estimates at a Mach number of 0.3 with an allowance for the losses observed in the UARL program, the LASRM inlet would have a dump recovery on the order of 0.95 or 0.96. The ACIM inlet with coaxial dump at a Mach number of 0.3 and an area ratio of 2.5 should have a dump recovery approaching 0.97.

In most offset inlet installations one or more sections of the inlet are physically separated from the vehicle body by a triangular device that diverts the vehicle boundary layer around the structure that houses the intersection of the inlet flow passage and the vehicle body wall. Three typical installations were shown in Fig. 7.4, and details of the vehicle boundary-layer diverter for an outward turning inlet (e.g., the ALVRJ) are shown in Fig. 7.12. The width w, as shown in the planform view, is established by the width of the structure at the inlet/vehicle body intersection. The length ℓ is set by the length of the inlet upstream from the leading forward edge of the intersection. Usually, the diverter length is less than the inlet length to preclude inlet ingestion of disturbances generated by the diverter shock/vehicle boundary-layer interactions. Width and length determine the wedge half-angle θ. The height h is selected as a function of the vehicle boundary-layer thickness δ at some operating condition in the mission trajectory, and the method for selecting h will be discussed subsequently. Height and width determine the diverter projected frontal area, which is shaded on the front view schematic in Fig. 7.12.

For an inward-turning inlet, such as the LASRM, the diverter is mounted upstream from the cowl lip (see Fig. 7.4) and is not contiguous with any element of the inlet. The width of the diverter is established as described earlier, but the length is relatively unconstrained. In order to prevent upwash of the flow from the sides onto the top of the diverter, a flat plate is mounted on the diverter as shown in the planform view in Fig. 7.13a. Leading edges of the plate are swept back along the shock waves generated by the diverter wedge. The side-mounted inlet requires a third type of diverter wherein the side wall is attached to the diverter. As shown in Fig. 7.13b, the diverter planform is a right triangle rather than the more typical isosceles triangle.

A product of vehicle boundary-layer exploration conducted during the decade of the 1960s in wind-tunnel tests of ogive/cylinder missile models

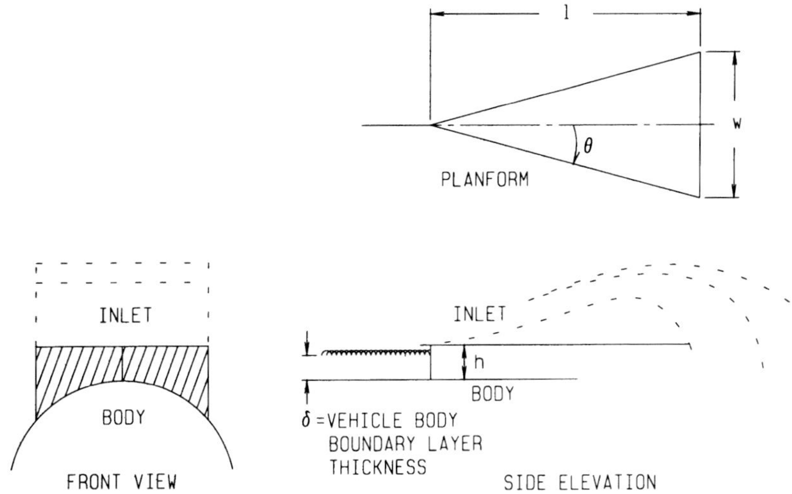

Fig. 7.12 Vehicle boundary-layer diverter details for an outward-turning inlet installation.

is a rule of thumb, which states that the vehicle boundary-layer thickness is 1% of the station at flight conditions of Mach 2.5 at sea level at a vehicle angle of attack of 0 deg. At vehicle station 50, therefore, the vehicle boundary-layer thickness δ should be 0.5 in., and at station 100, δ should be 1.0 in. Solving the flat-plate turbulent boundary-layer thickness equation presented in Chapter 4, i.e.,

$$\delta = \frac{0.377\ell}{(Re)^{\frac{1}{5}}} \tag{4.15}$$

yields values of 0.504 in. at station 50 and 0.876 in. at station 100. The good agreement between the values calculated at each station by the two methods may be a coincidence, but it seems to justify the use of Eq. (4.15)

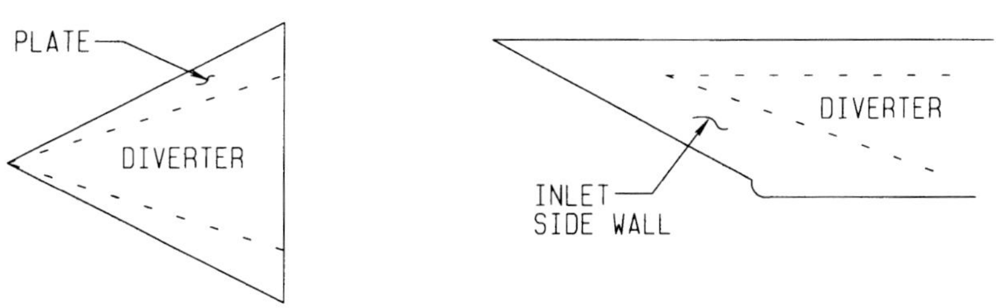

Fig. 7.13 Diverters for other inlet installations: a) inward-turning inlet; b) side-mounted inlet.

Table 7.3 Calculation of vehicle boundary-layer thickness at station 100 during operation at 0-deg angle of attack

Arbitrarily selected h, (ft)	M_∞	Reynolds no. per foot length Re/ℓ	Reynolds no. at station 100 $Re_{(\ell = 100'')}$	Eq. (4.15) $(Re)^{1/5}_{(\ell = 100)}$	$\backslash 0.377(8.333) =$	ft δ	in. δ	in. +14%
Sea level	2.0	142 × 10^5	1183 × 10^5	41.2	\ 3.1415	0.076	0.912	1.040
	2.5	178	1483	43.1		0.073	0.876	0.999
	3.0	213	1775	44.7		0.070	0.840	0.958
	3.5	249	2075	46.1		0.068	0.816	0.930
	4.0	285	2375	47.3		0.066	0.792	0.903
15,000	2.0	92	767	37.8		0.083	0.996	1.135
	2.5	115	958	39.5		0.080	0.960	1.094
	3.0	138	1150	40.9		0.077	0.924	1.053
	3.5	161	1342	42.2		0.074	0.888	1.012
	4.0	184	1533	43.4		0.072	0.864	0.985
30,000	2.0	57	475	34.3		0.092	1.104	1.259
	2.5	72	600	35.9		0.088	1.056	1.204
	3.0	86	717	37.2		0.084	1.008	1.149
	3.5	100	833	38.4		0.082	0.984	1.122
	4.0	114	950	39.4		0.080	0.960	1.094

45,000	2.0	31	258	30.4	0.103	1.236	1.409
	2.5	39	325	31.8	0.099	1.188	1.354
	3.0	46	383	32.9	0.095	1.140	1.300
	3.5	54	450	33.9	0.093	1.116	1.272
	4.0	62	517	34.9	0.090	1.080	1.231
60,000	2.0	14.5	121	26.1	0.120	1.440	1.642
	2.5	18	150	27.2	0.116	1.392	1.587
	3.0	22	183	28.3	0.111	1.332	1.518
	3.5	25.5	213	29.2	0.108	1.296	1.477
	4.0	29	242	30.0	0.105	1.260	1.436
75,000	2.0	7	58	22.5	0.140	1.680	1.915
	2.5	8.7	73	23.6	0.133	1.596	1.819
	3.0	10.5	88	24.5	0.128	1.536	1.751
	3.5	12.2	102	25.2	0.125	1.500	1.710
	4.0	14	117	25.9	0.121	1.452	1.655
90,000	2.0	4.4	37	20.6	0.153	1.836	2.093
	2.5	5.5	46	21.5	0.146	1.752	1.997
	3.0	6.6	55	22.3	0.141	1.692	1.929
	3.5	7.7	64	23.0	0.137	1.644	1.874
	4.0	8.8	73	23.6	0.133	1.596	1.819

to extrapolate results to other Mach number and altitude conditions. The equation has been solved at station 100 at freestream Mach numbers of 2.0, 2.5, 3.0, 3.5, and 4.0 at 15,000-ft increments in altitude between sea level and 90,000 ft. The calculations are shown on Table 7.3, where the boundary-layer thicknesses determined from Eq. (4.15) have been increased 14% to match the rule-of-thumb value at the Mach 2.5 sea-level condition.

The adjusted results are plotted in Fig. 7.14. At any given altitude, the vehicle boundary-layer thickness decreases with increasing freestream Mach number, and the variation in thickness between Mach 4.0 and 2.0 is less than 15% of the Mach 2.0 value. However, at any given freestream Mach number, the vehicle boundary-layer thickness increases between 10 and 15% for each 15,000-ft increase in altitude. The increase with altitude is nonlinear, but at every Mach number the boundary-layer thickness at 90,000 ft is just about double the sea-level value. It is postulated that these trends in boundary-layer thickness variation with freestream Mach number and altitude apply at any axial station on any ogive/cylinder body at 0-deg angle of attack. However, during vehicle pitch (or yaw) as the boundary-layer crossflows from the windward to the lee side, limited wind-tunnel data obtained at Mach 2.5 at a simulated altitude of 40,000 ft about two body diameters downstream from the ogive/cylinder junction of a particular configuration indicate a trend of 100% increase in lee side boundary-layer thickness for every 4-deg increase in angle of attack. It is expected that the windward side boundary-layer thickness would decrease a proportionate amount for every 4-deg increase in angle of attack. There were insufficient data to determine 1) if the increase (decrease) in vehicle boundary-layer thickness is linear with vehicle angle of attack; 2) how boundary-layer separation affects the trend on the lee side; 3) the variation of boundary-layer thickness with vehicle angle of attack at other stations; and 4) the effect of changes in operating Reynolds number on the boundary-layer thickness trends vs the angle of attack at any station.

Over the flight trajectory for a typical air-to-air mission, where the missile experiences a low-altitude launch at low speed, acceleration and climb to high speed, high-altitude cruise with a midcourse connection and a terminal maneuver, the vehicle boundary-layer thickness at any station will vary with time over a range of values similar in scope to those shown in Fig. 7.14. In order to preclude any vehicle boundary-layer ingestion by an inlet, the height of the diverter must equal the vehicle boundary-layer thickness at the highest value over the trajectory. At all other flight conditions, the diverter will be higher than necessary, and the drag contribution will be more than the minimum possible. It may be feasible to reduce the diverter height and allow inlet boundary-layer ingestion at one point in the trajectory with acceptable propulsion performance penalty. The acceptable limit on vehicle boundary-layer ingestion by any inlet is a function of all the variables discussed previously: inlet family, amount of compression, compression direction, and axial and circumferential position on the vehicle. The sensitivity of inlet performance to varying amounts of vehicle boundary-layer ingestion for a particular installation can only be determined in wind-tunnel tests with a specially designed model.

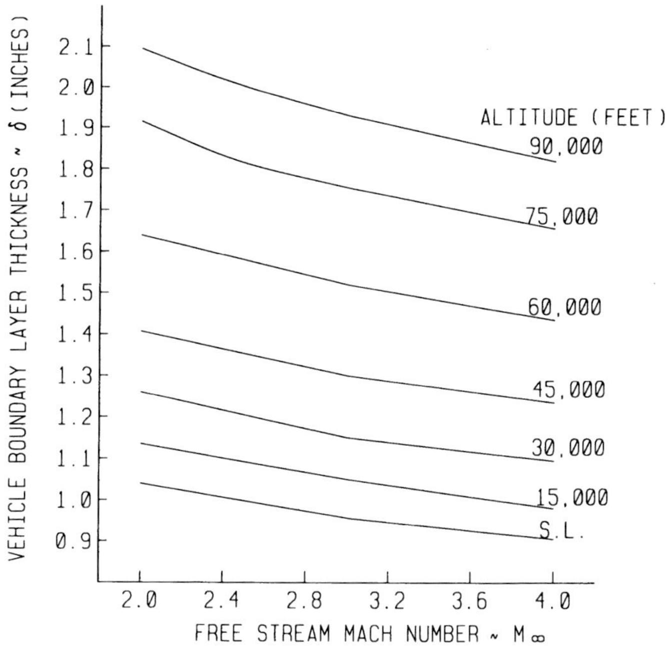

Fig. 7.14 Estimated variation of vehicle boundary-layer thickness at station 100 with freestream Mach number and altitude.

The technique for estimating the drag contribution of a vehicle boundary-layer diverter can best be explained with the aid of a hypothetical missile operating over a wide range of flight conditions similar to a typical air-to-air mission. The vehicle consists of a fineness ratio 2.25 Von Kármán ogive and a 15-in.-diam cylinder with an overall length of about 15 ft. The inlet is the Mach 3.0 design two-dimensional configuration shown in Fig. 3.8 and is mounted in the downward-compressing direction on the windward side of the vehicle with inlet ramp leading edge at station 100. The inlet has a square capture area that is 40% of the vehicle maximum cross-sectional area, or $A_C/A_3 = 0.40$, and $A_C = 70.69$ in.^{2} with height h = width w, = 8.41 in. The length of the inlet from the leading edge of the ramp to the intersection of the subsonic diffuser with the vehicle body wall is about 54.5 in. Thus, the boundary-layer diverter dimensions in the plan-form views, consistent with the notation in Fig. 7.12, are ℓ = 54 in., w = 8.4 in., and θ = 4.5 deg. The h is set at 2.0 in. to preclude any vehicle boundary-layer ingestion by the inlet at 90,000 ft altitude at a vehicle angle of attack of 0 deg, even though this condition may be only transitory. The boundary-layer diverter projected frontal area is approximately 23.8 in.^{2}, which is 34% of the inlet capture area A_C. The boundary-layer diverter drag coefficient is calculated in the same manner described in Chapter 6

Table 7.4 Calculation of diverter drag coefficient for a Mach 3.0 design inlet installed on a hypothetical missile as shown in sketch below

M_∞	M_0	Cp_0	×	$\frac{q_0/Pt_0}{q_\infty/Pt_0}$	×	$\frac{Pt_0}{Pt_\infty}$	+	$\left[\frac{P_0/Pt_0}{P_\infty/Pt_\infty}\right.$	×	$\frac{Pt_0}{Pt_\infty}$	$\left.-1\right]$	÷	$\frac{q_\infty}{P_\infty}$	=	Cp_∞	×	$\frac{dA}{A_C}$	=	C_{DD}
Initial 20%: two-dimensional pressure coefficient; deflection angle = 4.5 deg; $\frac{dA}{A_C} = 0.05$																			
2.0	2.0	0.100	×	1.0	×	0.987	+	1.0	×	0.987	− 1	÷	2.800	=	0.094	×	0.05	=	0.0047
2.5	2.5	0.082				0.977				0.977			4.375		0.075				0.0038
3.0	3.0	0.065				0.956				0.956			6.300		0.055				0.0028
3.5	3.5	0.058				0.932				0.932			8.575		0.046				0.0023
4.0	4.0	0.050		↓		0.904		↓		0.904			11.200		0.036		↓		0.0018
Aft 80%: conical pressure coefficient; deflection angle = 4.5 deg; $\frac{dA}{A_C} = 0.29$																			
2.0	2.0	0.030		1.0		S		1.0		S			S		0.025		0.29		0.0073
2.5	2.5	0.028				a				a			a		0.022				0.0064
3.0	3.0	0.025				m				m			m		0.017				0.0049
3.5	3.5	0.022				e				e			e		0.013				0.0037
4.0	4.0	0.020		↓				↓							0.010		↓		0.0028

M_∞	$C_{DD\ \mathrm{total}}$
2.0	0.0120
2.5	0.0102
3.0	0.0077
3.5	0.0060
4.0	0.0046

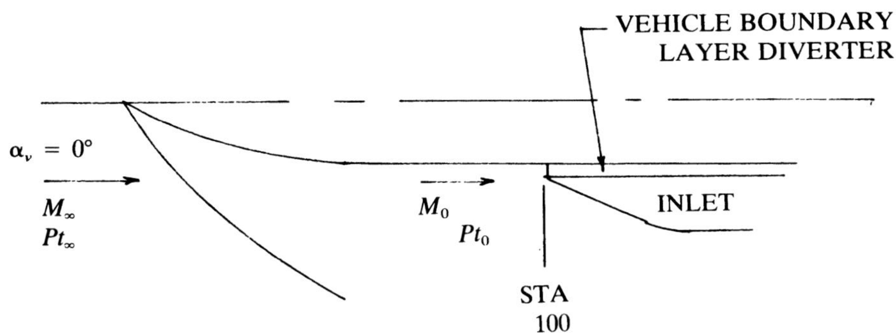

for pressure forces acting on forward-facing surfaces, i.e.,

$$C_{D_P} = C_{P_0}\frac{dA}{A_{\text{ref}}} \tag{6.12}$$

except that the pressure coefficient must be referenced to freestream (∞) conditions rather than local flowfield (0) conditions. The two pressure coefficients are related in accordance with the following:

$$C_{p_\infty} = \frac{C_{p_0}\dfrac{q_0}{Pt_0}}{\dfrac{q_\infty}{Pt_\infty}}\frac{Pt_0}{Pt_\infty} + \left[\frac{\dfrac{P_0}{Pt_0}\dfrac{Pt_0}{Pt_\infty}}{\dfrac{P_\infty}{Pt_\infty}} - 1\right]\Bigg/\frac{q_\infty}{P_\infty} \tag{7.9}$$

Pressure measurements obtained with static pressure taps along one wall of a typical vehicle boundary-layer diverter during exploratory tests conducted in the 1960s indicated that the flow remained two dimensional for about the initial 20% of diverter axial length; that is, the local pressure coefficients C_{p_0} are found in the two-dimensional curves of Ref. 1. Inasmuch as the flow area between the top surface of the inlet ramp and the vehicle body increases with increasing axial distance because of body curvature, the flow expands such that the average pressure coefficient over the aft 80% of the diverter approximates the conical values in Ref. 1. Recent measurements along the vehicle boundary-layer diverter in wind-tunnel tests of the ACIM configuration seem to confirm the 20% two-dimensional/80% conical flow process for estimating boundary-layer diverter drag coefficients. Application of Eq. (6.12) with the pressure coefficient correction of Eq. (7.9) to the boundary-layer diverter described above for the hypothetical missile yields the results given in Table 7.4.

The cowl lip, taper, and spillage drag coefficients calculated for the Mach 3.0 inlet and presented in Table 6.7 have been corrected in accordance with Eq. (7.9) in order to complete the installed drag coefficient picture. These corrected values and the diverter values given in Table 7.4 are plotted

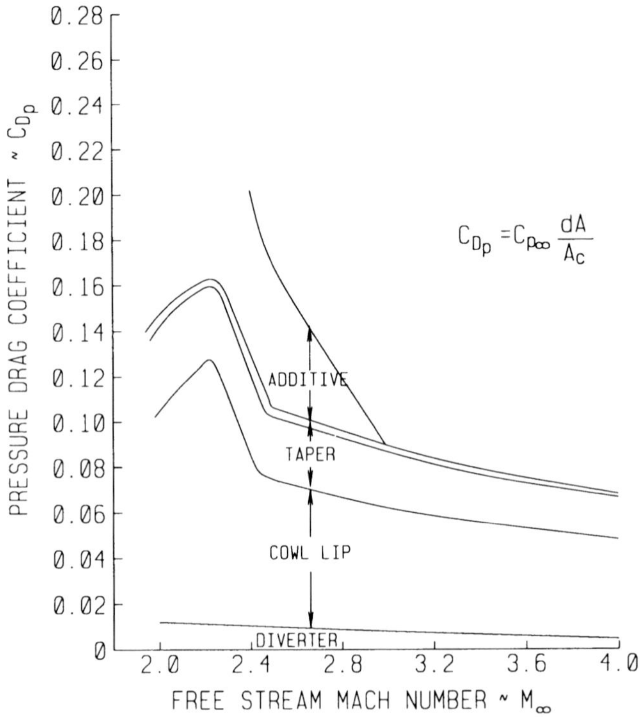

Fig. 7.15 Estimated variation of the pressure drag coefficient of an installed inlet designed at Mach 3.0 with freestream Mach number.

in Fig. 7.15. Even with the conservative design precept of no boundary-layer ingestion at any point in the trajectory, the vehicle boundary-layer diverter drag coefficient is less than 10% of the total installed inlet drag coefficient. However, if launcher box constraints required an inlet width to height greater than unity, both diverter frontal area and wedge half-angle would increase so that the diverter contribution to the total inlet pressure drag coefficient could be more than double that shown in Fig. 7.15 for the square inlet configuration. In the case of a high width-to-height ratio inlet, vehicle boundary-layer ingestion at some transient condition could prove less costly in terms of net thrust coefficient capability over the entire trajectory than the conservative no-ingestion approach.

8
VARIABLE GEOMETRY

8.1 Effects and Description

In Chapter 1 the subsonic combustion ramjet was compared to a closed channel with two throats, where the second throat, the nozzle, has a larger area than the first throat, the inlet, to accommodate increases in entropy and temperature between the throats. If the inlet throat is fixed at some percentage of the capture area (i.e., the inlet contraction ratio is determined at the inlet design Mach number), then the smallest possible nozzle throat area will be one that permits critical total pressure recovery at maximum heat release (equivalence ratio, $\phi = 1.0$). Thus, the engine is sized at the design Mach number of the inlet in accordance with the continuity [Eq. (1.25)], and the commensurate thrust coefficient, as determined from Eq. (1.17), is the maximum possible. Figure 8.1a shows a typical variation of thrust coefficient over a range of freestream Mach numbers for a fixed-geometry engine sized to produce maximum thrust at the inlet design point ($M = 3.0$) during operation at 0-deg angle of attack. The estimated variation of zero lift drag for a representative ramjet-sustained missile is also shown to illustrate the effective operating range where thrust equals or exceeds drag. Changes in the fixed-geometry engine thrust capability as a function of increased nozzle throat size with the same inlet are given in Fig. 8.1a for two larger nozzle throat areas. At and above the inlet design Mach number, the thrust capability is decreased because the inlet cannot attain the critical total pressure recovery level at $\phi = 1.0$. Below inlet design Mach number the larger nozzle throat sizes permit $\phi = 1.0$ engine operation to lower Mach numbers before throttling is required to preclude subcritical inlet operation.

If the nozzle throat area could be continually changed to maintain the engine thrust coefficient level at the left-hand break point of an infinite number of these curves, the engine thrust capability would be as shown in Fig. 8.1b. At and above the inlet design point, the engine would operate with the nozzle sized for maximum thrust as described earlier. Although the effective operating range of a missile with the variable-geometry nozzle engine would be increased only modestly (about 0.2 of a Mach number) compared with the maximum thrust fixed-geometry nozzle engine, the

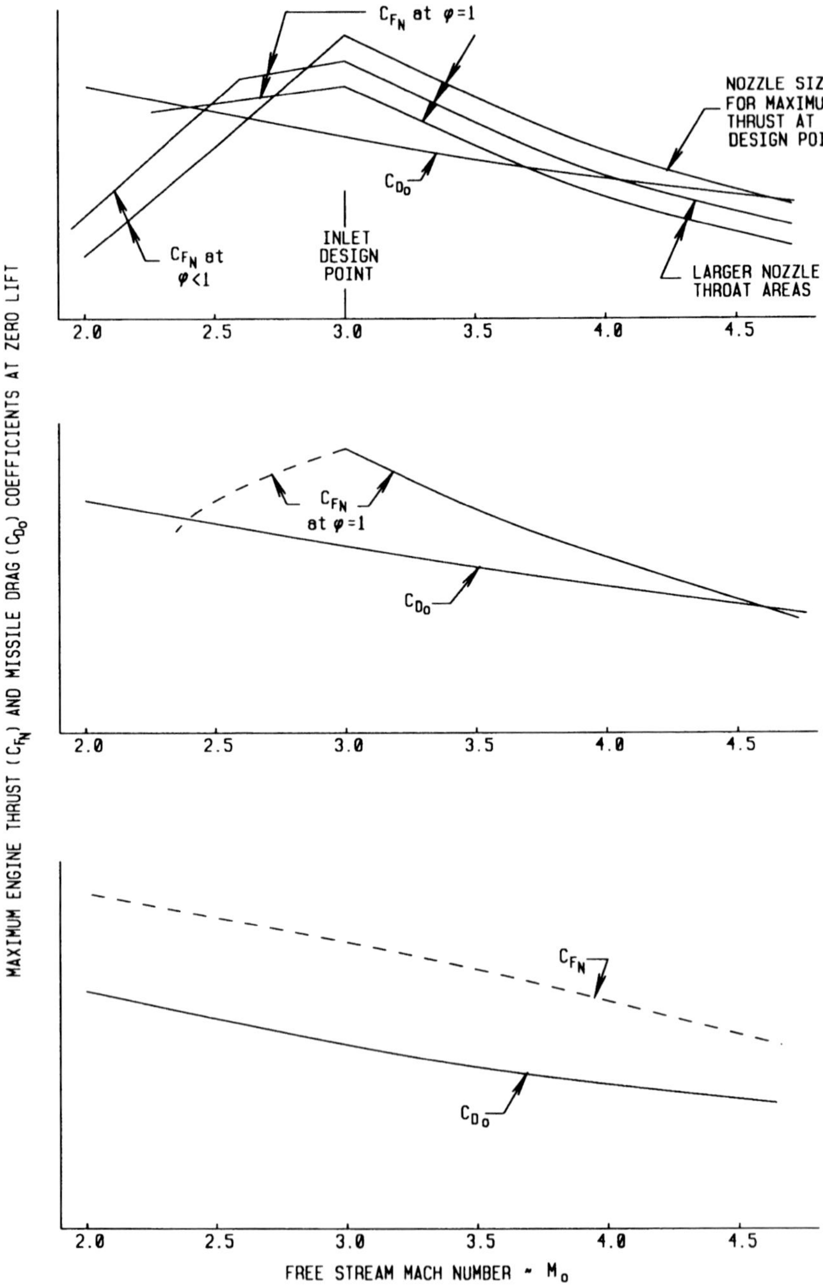

Fig. 8.1 Effect of component geometry on engine thrust capability: a) fixed-geometry inlet with three fixed-geometry nozzles; b) fixed-geometry inlet with variable-geometry nozzle; c) variable-geometry inlet with variable-geometry nozzle.

improvement in thrust capability occurs at a crucial flight condition where the ramjet sustain engine must take over from the burned out boost rocket.

If the inlet contraction ratio could be continually varied to maintain critical total pressure recovery at all freestream Mach numbers between two limits, such as 2.0–4.5, and the nozzle throat size could be independently varied to permit maximum heat release at the critical pressure, the engine thrust could exceed missile drag across the Mach number spectrum as shown in Fig. 8.1c. An engine with variable-geometry inlet and nozzle has the potential to extend the effective operating range of a missile over a much wider flight envelope than an engine with fixed-geometry components.

Several variable-geometry inlet configurations are shown in the sketches in Fig. 2.12 and are divided into two compression forms: all internal and mixed internal plus external. Although both forms have been investigated analytically and experimentally, only the mixed-compression form has been developed for operational use in flight. Generally, the flight inlets have a fixed cowl structure and a movable inner body, or ramp, which can be positioned to yield a large throat area for starting and smaller throat areas for running at different freestream Mach numbers over a prescribed range. In the running position at most Mach numbers, the internal compression is the primary contributor to the diffusion process. Because the amount of external compression is relatively low compared with an all-external-compression fixed-geometry inlet, the internal cowl surface can be aligned with the freestream flow direction, so that the cowl contribution to external drag is minimized.

For operation from takeoff across the subsonic and transonic regions and into the lower reaches of the supersonic region, thrust and mass flow matching requirements dictate variable-geometry engine components. The two-dimensional family of mixed-compression variable-geometry inlets designed for operation in this Mach number range is well represented by the Concorde intake, which is shown in the sketches in Fig. 8.2 reproduced from Ref. 26. The top sketch depicts the inlet positioned for takeoff with the hinged ramps set for maximum throat area and the auxiliary blow-in/blow-out door set for maximum air ingestion. The aft door in the bottom of the intake provides cooling air for the engine bay. The auxiliary door begins to close during climb, diminishing airflow into the engine, and closes completely at approximately Mach 0.7. From Mach 1.3 to supersonic cruise, the upper ramps are progressively lowered to the position shown in the middle sketch in Fig. 8.2. The isentropic compression forward ramp turns the flow so that an internal shock is generated by the misaligned cowl. Both the cowl shock and terminal normal shock (or train) are effectively trapped by the ramp bleed slot. Bleed air is used to cool the engine bay, and the bay door is closed. To prevent windmilling if an engine is shut down during supersonic flight, the upper ramps are lowered further as portrayed in the bottom sketch and the auxiliary door opens outward to spill air.

An example of the axisymmetric family of mixed-compression variable-geometry inlets for operation from takeoff to supersonic velocities above

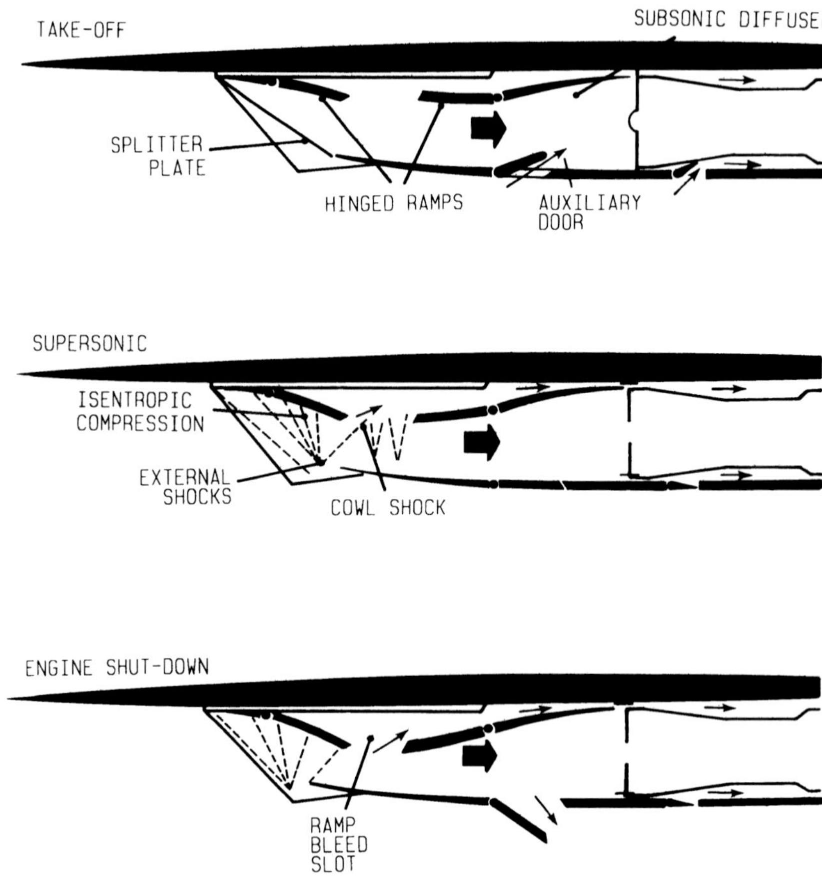

Fig. 8.2 Sketches of two-dimensional intake for the Concorde.

Mach 2.0 is given in Fig. 8.3, where an isometric view of the flight inlet of the YF-12 aircraft[27] illustrates the structural complexities involved in spike translation, inlet boundary-layer control, inlet unstart prevention, and inlet/engine airflow matching. The spike is set in the forward position for takeoff and subsonic operation and progressively retracted through the supersonic regime to the cruise condition. During cruise the shock trap bleed is in the throat region to mitigate adverse normal shock/boundary-layer interactions, and the forward bypass bleed is upstream from the throat. This latter bleed is part of the inlet unstart warning and rapid response control subsystem. The aft bypass door is opened or closed as required during flight to match inlet airflow to engine demands at off-design speed.

For operation across the upper end of the supersonic region and into the hypersonic regime (across a Mach number range of 3.0–8.0, for example) thrust and mass flow matching and boundary-layer control require-

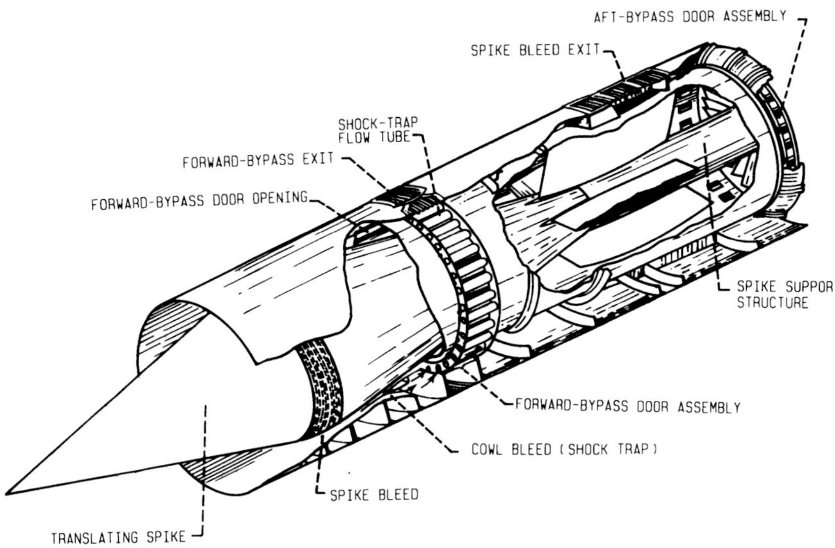

Fig. 8.3 Isometric view of flight inlet for YF-12 aircraft.

ments dictate variable-geometry engine components, which are much more complicated than their lower Mach number counterparts. However, before the components are described, it is necessary to define the operational environment. The boundary between supersonic and hypersonic flow is usually accepted as Mach 5.0.

At Mach 5.0 and above, air begins to depart rapidly from the ideal gas, where the ratio of specific heats γ equals 1.4. As shown in Ref. 1, a calorically imperfect gas requires different solutions to the equations of state, mass, momentum, and energy. Above about Mach 7.0 in the temperature range above 5000°R, gaseous imperfections such as molecular dissociation become important, and Mollier diagrams are mandatory for accurate solutions. Inasmuch as hypersonic flight trajectories will probably penetrate high altitudes where long runs of laminar boundary layer will be commonplace, the adverse pressure gradients associated with the inlet will induce early (near-leading-edge) separation and unacceptable inlet performance. A variety of inlets with differing amounts of adverse pressure gradient have been tested in many wind tunnels to determine a Reynolds number correlation that delineates a boundary between acceptable and unacceptable inlet performance. This boundary is shown on the bottom half of Fig. 8.4, where Reynolds number based on cowl diameter is plotted vs freestream Mach number. Inlets operating at Reynolds numbers below the boundary have critical total pressure recovery levels that are considerably less than inlets operating above the line. Although no visual proof is presented in the literature, laminar boundary-layer separation near the leading edges of compression surfaces is the probable cause of performance deficiencies. Of more importance from a practical standpoint is that the

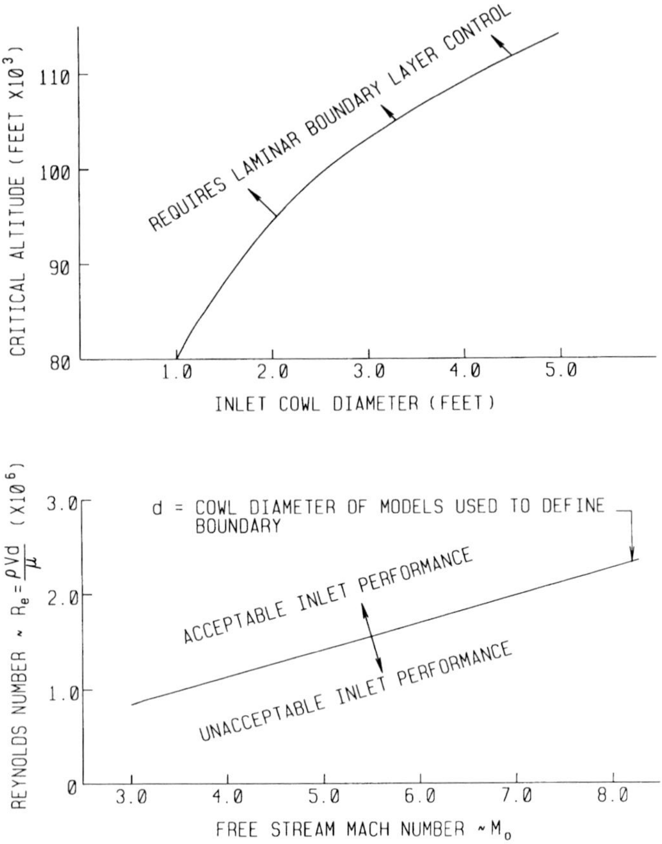

Fig. 8.4 Variation of critical altitude for laminar boundary-layer control with inlet cowl diameter based on a Reynolds number correlation determined from inlet wind-tunnel model test data.

boundary corresponds to a critical altitude that is a function only of inlet cowl diameter, as plotted in the upper half of Fig. 8.4. Operation above the critical altitude will require laminar boundary-layer control in the form of wall cooling, bleed, or trips, or some combination.

From the mid-1950s to the mid-1960s many government agencies sponsored research and development programs to investigate variable-geometry inlets for application to vehicles that could operate at hypersonic velocities. All of the investigators (Refs. 28–30, for example) learned that viscous affects were the primary source of total pressure loss in a hypersonic diffuser and that proper boundary-layer control techniques were the key to acceptable inlet performance. The schematic of a Mach 8.0 axisymmetric variable-geometry inlet in Fig. 8.5 is representative of the aerodynamic contours and length-to-diameter ratio of a hypersonic inlet and illustrates the exten-

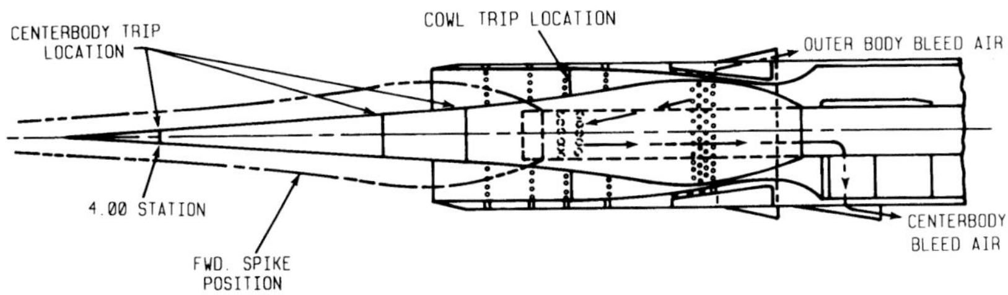

Fig. 8.5 Schematic of a Mach 8.0 axisymmetric variable-geometry inlet showing a representative boundary-layer control configuration.

sive boundary-layer control measures that were investigated during wind-tunnel model tests. Laminar boundary-layer control included trips of various height on both spike and cowl with forward cowl bleeds to help prevent separation. Bleeds on both compression surfaces in the throat region were designed to control turbulent boundary-layer separation. Many variations in the amount and location of the initial boundary-layer control measures shown in Fig. 8.5 were tested. The centerbody could be translated manually over a wide range of positions to determine the starting position at each freestream Mach number investigated and the running position as a function of boundary-layer control changes. No provisions were made for automatic spike position control or prevention of accidental unstart. The model was uncooled, but the flight environment will dictate wall cooling for structural integrity.

8.2 Hypersonic-Design Procedures

Operating Goals

The measured running, or operating, contraction ratios for a variety of all internal and mixed internal plus external compression inlets were shown in Fig. 2.14 over a range of supersonic freestream Mach numbers between 2.0 and 4.0. This curve can be approximated by an isentropic process wherein the flow is compressed to a throat Mach number equal to one-half of the freestream Mach number. Extrapolation of the isentropic approximation into the hypersonic regime yields the contraction ratio curve shown on the bottom half of Fig. 8.6. The corresponding critical total pressure recovery is calculated from the product of a normal shock recovery at one-half of freestream Mach number and the empirically determined subsonic diffuser recovery value of 0.9, as discussed in Chapter 3. Critical total pressure recoveries and commensurate kinetic energy efficiencies are plotted in the top half of Fig. 8.6 over the freestream Mach number range of 3.0–8.0. The parameters shown in conjunction with the assumption of full capture across most of the Mach number range and low drag compared with a fixed-geometry inlet of comparable size form the hypersonic variable-geometry inlet operational performance goals.

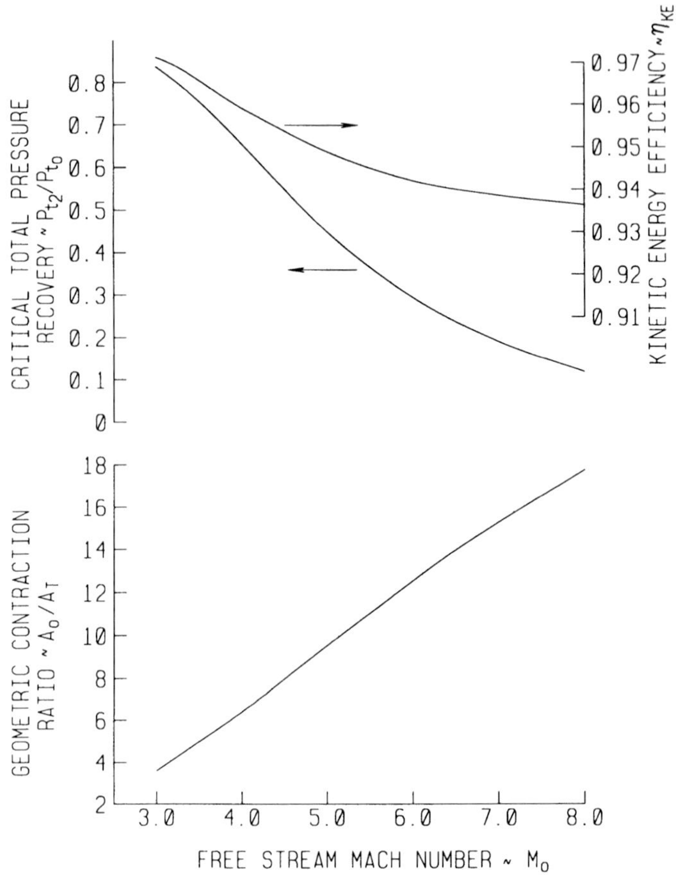

Fig. 8.6 Hypersonic variable-geometry inlet operational performance goals.

Design Point

Inasmuch as aerodynamic contours can be established only at one Mach number, it is unlikely that any one inlet can achieve the performance goals across the Mach number range of 3.0–8.0. For each freestream Mach number there is an optimum length compatible with best performance, and this length increases with increasing freestream Mach number. An inlet designed at Mach 8.0 would be too long (excessive viscous losses generated by too much wetted surface) for efficient operation at Mach 3.0. Conversely, a Mach 3.0 inlet would be much too short (very high surface pressure gradients promoting boundary-layer separation) for efficient operation at Mach 8.0. Tests with inlets designed at Mach numbers of 4.7, 5.5, and 8.0 demonstrated that the best performance across the widest part of the Mach 3.0 to 8.0 range was achieved with the Mach 8.0 design inlet.

Potential Flow

Although viscous effects are predominant in setting achievable performance levels for any hypersonic inlet, contour development must be predicated on the best potential flow design practices. A well-designed hypersonic diffuser should compress the freestream flow isentropically to an undistorted flow profile at the inlet throat. Contours should be defined to preclude coalescence of compression waves or strong shocks that engender abrupt changes in the pressure gradient along the diffuser surfaces. The progression of events in the development of aerodynamic contours for a Mach 8.0 design axisymmetric inlet with a translatable innerbody is highlighted in Fig. 8.7, where three sketches of the top half of a skeletal inlet are drawn. From the top the first element is the cowl surface, which is a portion of a stream line lifted from a two-dimensional characteristics net, as described in Chapter 2. The Mach line from the cowl lip to the innerbody centerline defines the upstream boundary where the freestream is uniform and parallel at Mach 8.0. At the downstream end of the cowl, the boundary condition is an undistorted throat flow at Mach 4.0. The rate of change of Mach number (pressure gradient) along the cowl surface from the lip to the throat station can be arbitrarily specified and represented by a mathematical expression. Cowl coordinates, several different Mach number distributions along the cowl, and the boundary conditions were programmed into a computer-generated axisymmetric characteristics net to determine the inner-body contour shown as a dashed line. Several iterations were required to develop a satisfactory inner body that complemented the cowl in compressing the flow isentropically to an undistorted throat profile. Inasmuch as the full isentropic spike tip developed by the machine is structurally impractical, a 4-deg half-angle conical tip is fitted to the inner body as shown in the middle sketch. The oblique shock generated by the conical tip impinges on the cowl lip at the Mach 6.4 running position. At the Mach 8.0 running position, the shock is downstream from the cowl lip as shown on the sketch. In consonance with the design practices discussed

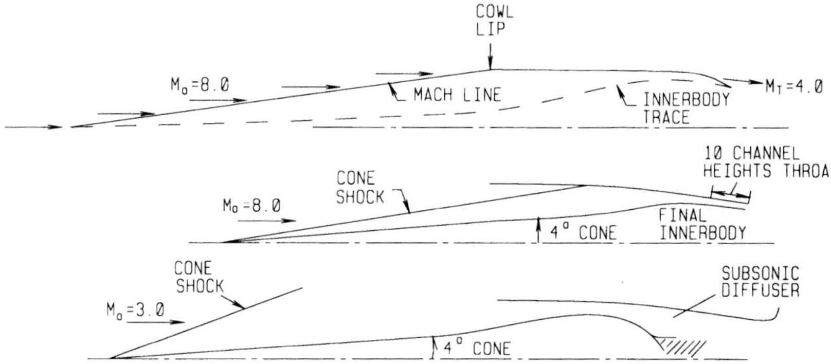

Fig. 8.7 Contour development for a Mach 8.0 variable-geometry hypersonic axisymmetric inlet.

in Chapter 3, a throat length of 10 channel heights is incorporated at the Mach 8.0 running position (middle sketch), and subsonic diffuser lines are defined at the Mach 3.0 running position (bottom sketch).

Viscous Effects

All contours produced in accordance with the prescribed potential flow practices were tested against the pressure gradient parameter, $(\mathrm{d}P/\mathrm{d}x)(x/q)$, given in Fig. 4.7 to determine that the gradients are within safe limits for turbulent boundary-layer attachment. Furthermore, the gradients engendered by the 4-deg cone shock sweeping across the cowl surface are within safe limits. Near the throat region, however, where the boundary layers from the two compression surfaces begin to merge and eliminate the potential flow core, the pressure gradient parameter probably has no practical significance. Prudence dictates that more positive boundary-layer control procedures, such as bleed, or cooling, be applied. In the development of any hypersonic inlet, provisions should be made for bleed in and upstream from the throat on both the inner body and the outer body in conjunction with cooling passages. Generally, flush holes in several rows should be located in the region of highest pressure gradients. Because of the possibility of operating above the critical altitude for laminar boundary-layer separation, trips of various heights for both spike and cowl and forward cowl bleed should also be provided.

Real Gas

By means of the charts and graphs presented in Ref. 1, it is possible to compute real gas pressures, temperatures, and other flow parameters for any geometric shape at any freestream Mach number and altitude condition. Above Mach 4.0 at any altitude, real-gas stagnation pressures are higher than perfect gas values, whereas real-gas stagnation temperatures are lower. For example, an engine operating at Mach 9.0 at 82,000 ft altitude would experience a real-gas stagnation pressure about twice the perfect-gas value (150 psi vs 75 psi) and a real-gas stagnation temperature on the order of 5800°R compared with 6900°R under perfect-gas conditions. Thus, real-gas flow will affect hypersonic engine structural design and cooling requirements. Oblique shock relationships are also changed by real-gas effects, but for two-dimensional deflection angles less than 10 deg the effects are negligible. For the hypersonic inlet configurations described herein, the flowfield and the contours along most of the hypersonic diffuser upstream from the throat region should be the same under real-gas conditions as under perfect-gas conditions. Normal shock recovery is usually lower with real-gas flow than with perfect-gas flow, but the difference diminishes as the throat Mach number decreases. In order to minimize the effect of real gas on inlet performance, the flow should be compressed to a low throat Mach number, such as 2.0 or below. For a low throat Mach number, the throat area of the inlet must be considerably less than the potential flow design value, and the increased contraction ratio will increase the surface pressure gradients well above the potential flow safe levels.

The behavior of a boundary layer navigating high pressure gradients in a real-gas environment with realistic heat-transfer rates is currently undocumented. Until evidence is presented that real gas has more than a second-order effect on hypersonic inlet contours and performance, it is recommended that no corrections be applied.

Design Completion

The design procedures described earlier have been incorporated into the inlet model presented in Fig. 8.5. Except for differences in the number and location of bleed holes, a flight inlet should look exactly like the model that was fabricated for testing in the JPL 20-in. supersonic and 21-in. hypersonic wind tunnels. The cowl lip is a 10-deg included-angle wedge that is designed to ensure structural rigidity near the stagnation point in any flight situation. For model operation of up to 5-deg angle of attack in the two tunnels, the maximum permissible cowl diameter is 5 in. Because of the 4-deg half-angle cone tip on the spike, the capture area of the inlet is less than unity at the operating contraction ratios shown in Fig. 8.6 for all freestream Mach numbers less than 6.4. Capture area ratios across the Mach number range of 3.0–8.0 are plotted in Fig. 8.8, and the method of calculating values with the spike cone shock upstream from the cowl lip is illustrated by the sketch and equations at the top of the figure. These capture area ratios represent the mass flow crossing the cowl lip plane and, in the absence of any bleed flow, the mass flow through the model plenum or the flight engine combustion chamber. Because of the flat inner cowl surface, the cowl wave drag coefficient is determined from the product of the pressure coefficient for a 10-deg wedge and the projected frontal area of the wedge referenced to the capture area. In contrast, an all-external compression fixed-geometry inlet with inner cowl surface aligned with the local flow direction has outer cowl surface angles on the order of 30–40 deg depending on the amount of flow turning specified in the design. If the modest compression Mach 3.0 design example inlet of Chapter 3 with an external cowl surface angle of 28.5 deg and the Mach 8.0 variable-geometry inlet were deliberately designed to have the same ratio of projected cowl lip frontal area to capture area, the total cowl wave drag of the Mach 3.0 inlet would be more than seven times higher than the cowl wave drag of the Mach 8.0 inlet at a freestream Mach number of 4.0. (The estimated wave drag of the Mach 3.0 inlet over a Mach number range of 2.0–4.0 is given in Fig. 6.17.) A similar ratio would apply at Mach numbers between 4.0 and 8.0. At Mach numbers below 6.4, where the Mach 8.0 variable-geometry inlet spills air, there is an additive drag penalty, but the spillage occurs at such a low flow angle (1 deg or less behind the 4-deg conical shock compared with 15–20 deg for a modest compression fixed inlet) that the drag increase is very small until the isentropic compression fan moves upstream from the cowl lip at a freestream Mach number of about 3.5. Thus, the variable-geometry inlet without any bleed has a significant wave drag advantage over a fixed-geometry inlet with only modest total pressure recovery capability.

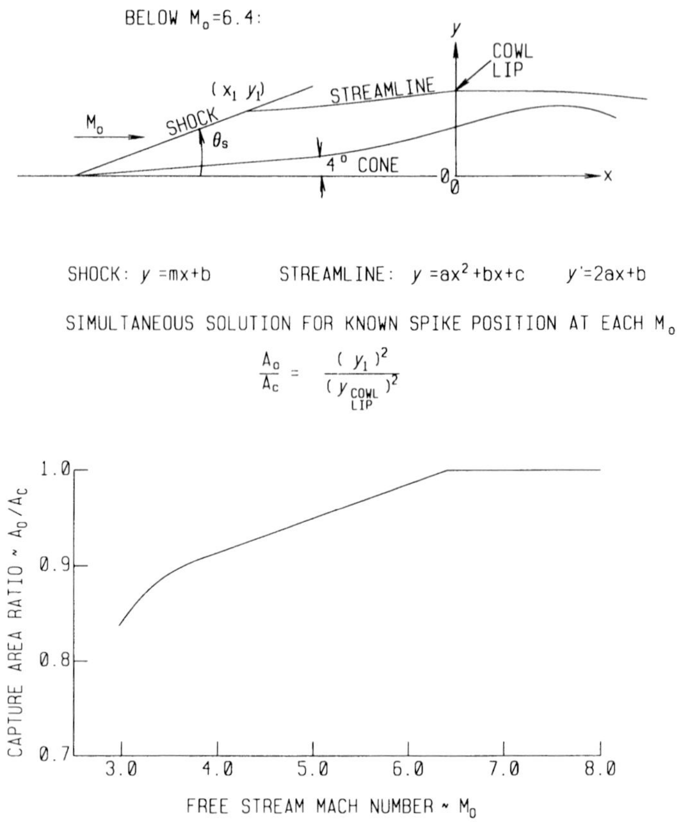

Fig. 8.8 Estimated capture area ratios of Mach 8.0 variable-geometry axisymmetric inlet at maximum contraction ratio at each freestream Mach number.

Hardware Variables

Although the aerodynamic contours and structural design features remained unchanged through three series of tests in the two wind tunnels, the boundary-layer control configuration underwent about a dozen revisions. Several times between runs in the initial tests, modifications, such as enlarging or plugging existing bleed holes or drilling additional rows of holes, were accomplished at the tunnel facility. Between each test series extensive modifications were completed after a thorough analysis of the preceding test results. The last modification completed before the third test series produced the final configuration shown in Fig. 8.9. A new inner body with 12 rows of 0.10-in.-diam holes was constructed for use during the supersonic tests. The bleed holes are located in the region where the internal cowl lip shock generated by the misalignment between the capture stream tube and cowl at the lower Mach numbers impinges on the inner body. The holes were initially filled with dental plaster, which could be drilled out to change location from forward to center to aft up to a maximum

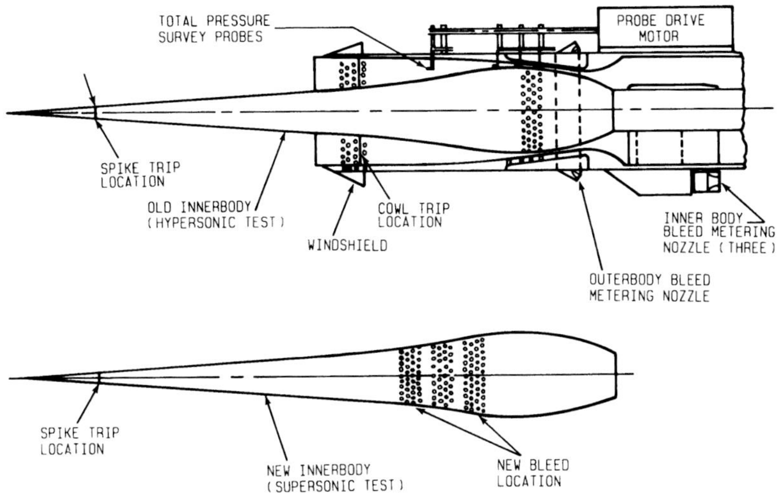

Fig. 8.9 Final test configuration of the Mach 8.0 axisymmetric variable-geometry inlet model.

of four rows open at any one time. The maximum allowable bleed flow rate is limited by the size of the annulus between the inner-body drive shaft and the tunnel model support sting. There were no throat bleeds in the inner body for the supersonic tests. The four rows of bleed holes in the outer body common to both tests were closed in both tunnels. The cowl trip location, shown at a station 5 in. downstream from the cowl lip in Fig. 8.5, was moved to a station 2 in. from the lip, and four rows of 0.10-in.-diam holes were drilled in the forward cowl (three upstream and one downstream from the trip station). These holes remained open during all tests in both tunnels. The hypersonic inner body had four rows of 0.040-in.-diam holes in the throat region which remained open during all hypersonic testing. Both the inner-body and outer-body throat bleed were collected in plenums and discharged through metering nozzles. The forward cowl bleed was dumped directly overboard. Four total pressure survey probes, spring-loaded and driven by cams attached to a common shaft rotated by a remotely operated electric motor, were assembled and mounted on the aft section of the outer body. At any operating position of the inner body, the probe closest to the throat could be activated to measure flow properties across the annulus. Because the probe mechanism separated the external cowl boundary layer and affected the cowl bleed, the windshield was tack welded to the cowl lip wedge.

8.3 Hypersonic-Performance Capability

General

The 5-in.-diam. Mach 8.0 design variable-geometry inlet model with the dozen variations in boundary-layer control configuration was investigated in three series of tests in the JPL 20-in. supersonic and 21-in. hypersonic tunnels at freestream Mach numbers of 2.61, 3.01, 3.99, 4.54, 5.01, 6.06, 7.35, 8.56, and 10.34 and at angles of attack of 0 and 5 deg. Most of the performance values presented herein were acquired with the boundary-layer control configuration shown in Fig. 8.9 and described in the preceding section. Because of limitations on the test time available in the two tunnels, no one boundary-layer control configuration was tested at every Mach number; hence, the results are incomplete. Furthermore, the performance results are not necessarily the best that can be achieved with this type inlet at any given Mach number, and the associated boundary-layer control configuration at each Mach number is not necessarily the optimum that can be applied. Although the Mach number range of 3.0–8.0 is of primary interest in this text, some performance results obtained at Mach 8.56 are presented to show the effect of operating a variable-geometry inlet above design Mach number.

Overall Inlet

Inlet kinetic energy efficiencies and the commensurate geometric contraction ratios are shown in Fig. 8.10 in comparison with the target values previously defined between the freestream Mach numbers of 3.0 and 8.0. Although the measured contraction area ratios equal or exceed the target numbers across the Mach number range of interest, the kinetic energy efficiencies approach the target only near the upper Mach number limit and deviate at an increasing rate with decreasing test Mach number. In the hypersonic region between 5.0 and 8.0, the kinetic energy efficiencies lie between 0.935 and 0.938, which is well above the best performance ($n_{kE} \sim 0.92$) that could be expected with a conservatively designed fixed-geometry hypersonic inlet operating at its design point. At Mach 3.0, however, the measured n_{kE} is less than could be achieved with the conservatively designed fixed inlet (refer to Fig. 6.24). Determination of the n_{kE} level at Mach 8.0 was accomplished by extrapolating the curves from the neighboring data points on semilog paper, using an above design total pressure recovery vs Mach number slope commensurate with 10 channel heights of throat (see Fig. 6.4). There are insufficient data to conclude that variable-geometry inlets follow the same exponential rule above design Mach number as fixed inlets, but the available data fit the requisite pattern. Fairing of the contraction ratio curve is arbitrary and is shown dashed to highlight that fact.

Capture area ratios, which represent the total mass flow crossing the cowl lip plane and are calculated from the sum of the measured flow through the model plenum, the measured flow through the throat bleeds, and estimated flow through the forward cowl and inner-body bleeds, are compared in the top section of Fig. 8.11 with the target curve. Generally, the agreement is well within the limits of experimental accuracy. Bleed flow rates shown in the middle section of Fig. 8.11 are the result of different bleed configurations tested in the supersonic and hypersonic tunnels. Supersonic tests were conducted with forward bleeds only, whereas hyper-

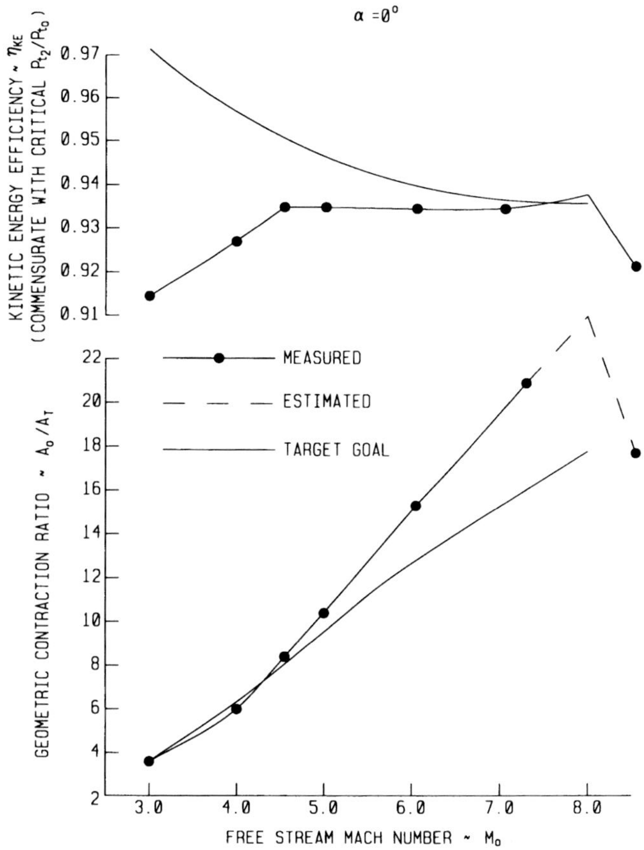

Fig. 8.10 Mach 8.0 variable-geometry axisymmetric inlet overall performance.

sonic tests featured inner-body throat bleeds as well as forward cowl bleeds. Inasmuch as the driving pressure in the forward region of the inlet is so much less than the throat pressure, the mass flow through the forward bleeds is always much less than the throat bleed flow rate. The increase in bleed flow rate below Mach 4.0 reflects the increasing strength of the internal cowl lip shock as the captured streamline approaches the cowl lip at increasing angles of attack. If it is assured that no bleed momentum is recovered and Eq. (6.18) applies, the drag coefficient chargeable to the bleed is given in the bottom of Fig. 8.11. In the supersonic region the bleed drag coefficient is about double the cowl lip wave drag coefficient, but the total drag coefficient at Mach 4.0 is still only about one-fourth the drag coefficient of the Mach 3.0 design fixed-geometry inlet. However, in the hypersonic region the bleed drag coefficient in the worst-case scenario of no momentum recovery is an order of magnitude greater than the cowl lip wave drag coefficient, and the total drag coefficient of the variable-

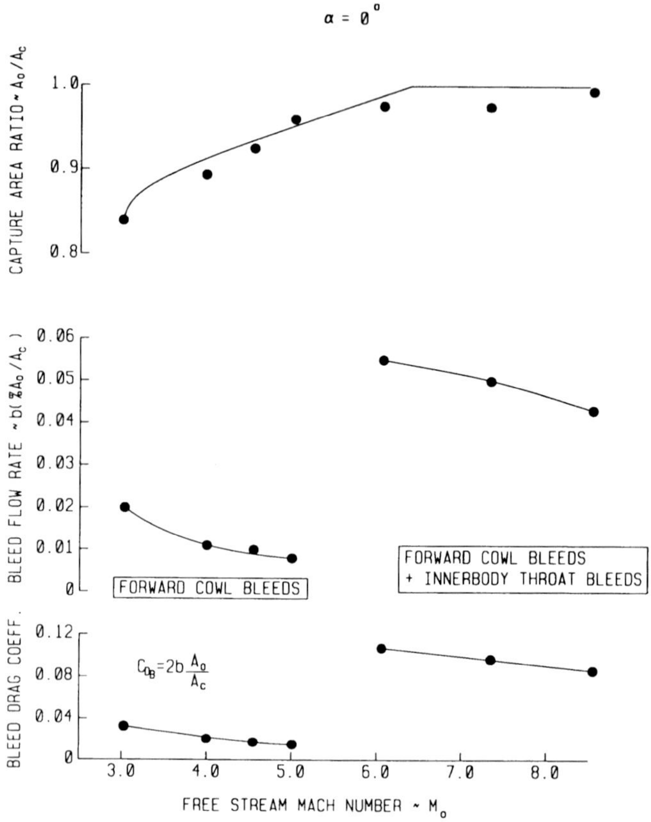

Fig. 8.11 Mach 8.0 variable-geometry axisymmetric inlet overall performance.

geometry inlet would be greater than the fixed-geometry inlet. If the throat bleed plenum and nozzle could be designed to achieve a kinetic energy efficiency of only 0.10–0.15, the total drag coefficient of the variable inlet would be on the same order as the fixed inlet operating at hypersonic speeds. Opening all outer-body throat bleeds more than doubled the bleed flow rates in the hypersonic region, but showed no improvement in total pressure recovery. There was insufficient test time to determine a minimum throat bleed flow rate consistent with the highest recovery levels achieved.

Hypersonic Diffuser

In order to isolate the performance of the diffuser, at the operating position of the inner body at each freestream Mach number, the survey probe closest to the upstream end of the throat was activated. Flow properties at selected radii across the annulus were determined from the survey probe readings, wall static pressures at the survey stations, and the Rayleigh pitot formula. Throat Mach number and total pressure recovery profiles change from a relatively full type with a central core unaffected by viscosity at Mach 3.0 to a parabolic-type characteristic of fully developed pipe flow at Mach 7.35. At the intermediate Mach number of 5.01, although the profile is parabolic with very low total pressure recovery near the walls, there is a significant central core flow with very high total pressure recovery. The extreme variation of flow conditions across the passage indicates that no arbitrarily selected stream tube can represent the average flow properties at the inlet throat station. Both the mass and momentum of the highly distorted profiles can be reproduced by the averaging method of Ref. 31 to yield uniform one-dimensional profiles that can be used to define reasonably realistic performance values.

One-dimensional values of total pressure recovery across the diffuser and of throat Mach number are given in Fig. 8.12. The total pressure recovery levels reflect the magnitude of the losses, which are primarily viscous but also include oblique shock and mixing, between freestream

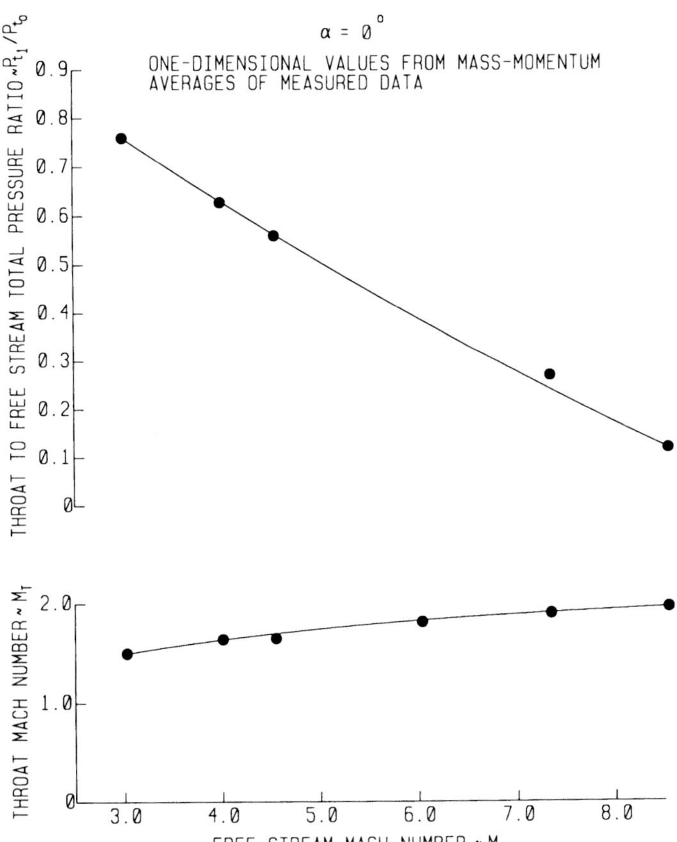

Fig. 8.12 One-dimensional values from mass-momentum averages of measured data.

and the throat station of the inlet. At Mach 8.0, for example, only about 20% of the total pressure head available in the freestream has been recovered. At the lower Mach numbers the forward inner-body bleed did mitigate cowl lip shock-induced losses compared with no bleed in previous tests, but had no effect on viscous losses. Despite the losses, the one-dimensional throat Mach numbers indicate a degree of flow compression that is generally higher than the potential flow design value at all freestream Mach numbers. It seems obvious that boundary layers, when properly controlled, play a major role in the hypersonic compression process. Throat Mach numbers less than 2.0 should also minimize the effects of real gas on normal shock pressure recovery.

The one-dimensional values presented in Fig. 8.12 furnish an alternate method for calculating overall inlet total pressure recovery (i.e., the product of the hypersonic diffuser recovery, Pt_1/Pt_0, normal shock recovery at the throat Mach number, M_T, and the empirical subsonic diffuser recovery of 0.90). This one-dimensional set, the measured critical total pressure recoveries corresponding to the kinetic energy efficiencies in Fig. 8.10, and the target potential flow approximation are plotted vs freestream Mach number in Fig. 8.13. All three curves are practically identical in the Mach number range between 6.0 and 8.0. Below Mach 6.0, the target curve deviates from the other two, and the deviation increases with decreasing Mach number. Generally, the one-dimensional and measured values agree within about 5%, and the difference between these two curves and the potential flow target at the lower Mach numbers is a measure of the effect of viscous losses on overall total pressure recovery with an inlet that is longer than optimum. To illustrate the validity of the potential flow target at low Mach numbers, the measured total pressure recovery of a Mach 3.0 design variable-geometry inlet with bleed flow rates on the order of 4.5% (Ref. 32) has been included in Fig. 8.13. At Mach 3.0, the measured performance of the inlet designed for optimum performance at that Mach number matches almost exactly the potential flow target. Inasmuch as the Mach 8.0 design inlet is about five times longer than the Mach 3.0 design, the sizable difference in wetted surface area should be expected to generate substantial friction losses. The measured critical total pressure recovery of a Mach 8.0 design, variable-geometry, water-cooled, copper inlet model with no bleed[30] has also been added to Fig. 8.13 to indicate that wall cooling is as effective as bleed in the attainment of the target performance at Mach 8.0. Both the water-cooled Mach 8.0 and the Mach 3.0 variable-geometry inlets deviate from the potential flow target at Mach numbers below the respective design values. Available data confirm that the potential flow approximation is an adequate representation of performance wherever inlet contours and lengths are optimum.

Laminar Boundary-Layer Control

As discussed in Chapter 4, a natural transition from laminar to turbulent boundary-layer flow occurs at a critical Reynolds number that is a function of a host of variables. The arbitrary average of many observed values of

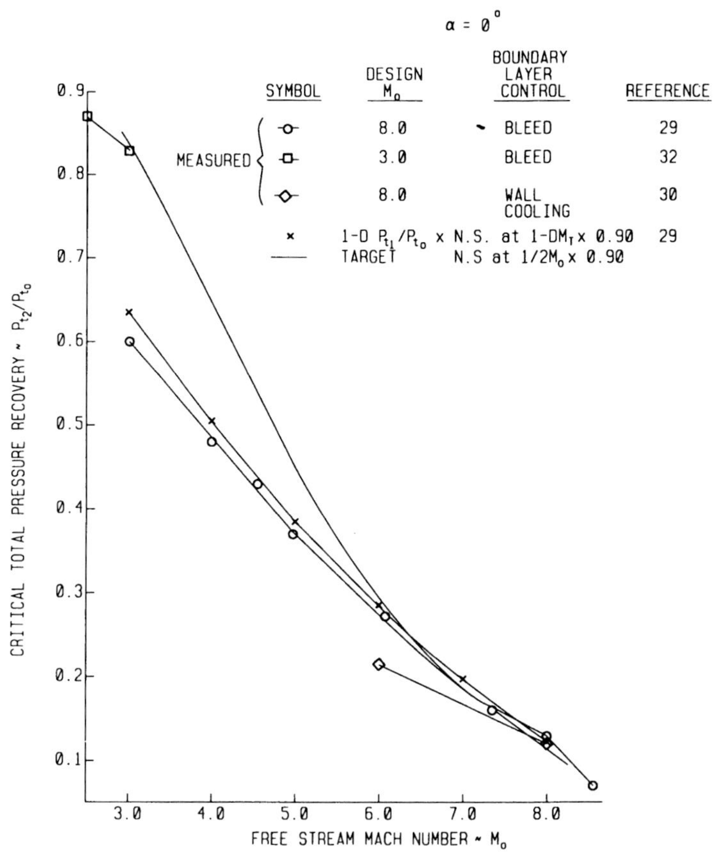

Fig. 8.13 Critical total pressure recovery of three variable-geometry inlets.

critical Reynolds number as plotted in Fig. 4.4 indicates that the transition occurs at increasingly greater distances from boundary-layer inception as the Mach number increases across the hypersonic regime; i.e., ever-longer runs of stable laminar flow are promoted with increasing hypersonic Mach numbers. The Reynolds number correlation shown in Fig. 8.4 implies that the adverse pressure gradient associated with an inlet can, under certain conditions, induce laminar boundary-layer separation before natural transition can occur. Operation at altitudes above a critical level, which is a function of only inlet cowl diameter, seems to be a sufficient condition to cause the laminar separation problem, and control measures, such as trips and/or bleed and/or wall cooling, are required. Inasmuch as the 5-in.-diam Mach 8.0 inlet model operating at the maximum Reynolds number available in the 21-in. hypersonic tunnel would be simulating altitudes above the critical value for all test Mach numbers above about 5.8, provisions were made for both bleed and trips. Unfortunately, the critical altitude curve in Fig. 8.4 does not specify the proper amount or location of boundary-layer control; hence, successful control was achieved in a series of arbitrary steps.

At Mach numbers of 6.06 and above in the initial test series, both spike and cowl boundary layers separated, as determined from axial static pressure measurements, and maximum contraction ratios of only 4.0–5.0 could be achieved before inlet unstart. At all Mach numbers, the spike boundary-layer separation problem was cured with a 0.040-in.-high trip, and this trip had no significant adverse effect on performance at any lower Mach number, where the trip was not required. No cowl boundary-layer separation was observed at Mach 5.05 (operation below the critical altitude), but the model was tested with three rows of 0.040-in.-diam holes (72 per row) open in the forward cowl region, as shown in Fig. 8.5. At Mach 6.06, the same forward cowl hole pattern was insufficient to ensure attainment of the target contraction ratio with concomitant desirable performance, and the first two rows of holes were enlarged to 0.080-in. diam. This modification proved successful, and no cowl trip was required. The cycle was repeated at Mach 7.34, where successful operation was achieved with the first two rows of holes enlarged to 0.10-in. diam and two additional rows of 0.10-in.-diam holes drilled on either side of the original first row. No cowl trip was needed at Mach 7.34. The forward cowl bleed area variation with hypersonic freestream Mach number during the evolutionary tests with the trip station 5-in. from the cowl lip is summarized on the upper half of Fig. 8.14. The curve labeled "sufficient bleed" connects the bleed areas in the forward cowl that permitted achievement of the target contraction ratio at each freestream Mach number. These are not necessarily the minimum bleed areas. The insufficient bleed curve shows that successful operation was not achieved at Mach 8.56 despite the presence of up to 12 rows of 0.10-in.-diam holes and the insertion of several different height trips. Analysis of the data revealed that the 5-in. cowl trip station was located well downstream from the initial boundary-layer separation point, and no amount of control in the wrong location could solve the problem. The trip station should be located in a region of relatively low pressure gradient but sufficiently far aft from the leading edge that transition can be forced. A trip station Reynolds number of 300,000 based on distance from cowl lip to trip station has been used as the criterion to relocate the trips as shown in the final configuration (Fig. 8.9). The 2-in. trip station in conjunction with the four rows of 0.10-in.-diam forward cowl bleed holes permitted successful operation at Mach numbers of 5.05, 6.06, and 7.35 without cowl trips and at Mach 8.56 with two different trip heights.

The effect of trip height variations on the kinetic energy efficiency of the final configuration at Mach 8.56 is shown in the bottom half of Fig. 8.14. With no cowl trip the n_{kE} is only 0.82, which is not far above the normal shock value, which is also shown. For a range of trip heights of 0.09–0.114 in., the kinetic energy recovery is almost constant at 0.92, which is lower than the potential flow target, but is consistent with the level that should be expected for an inlet operating above design Mach number. Trip heights lower than the "optimum" range are ineffective in forcing transition. Trip heights higher than those required protrude through the bound-

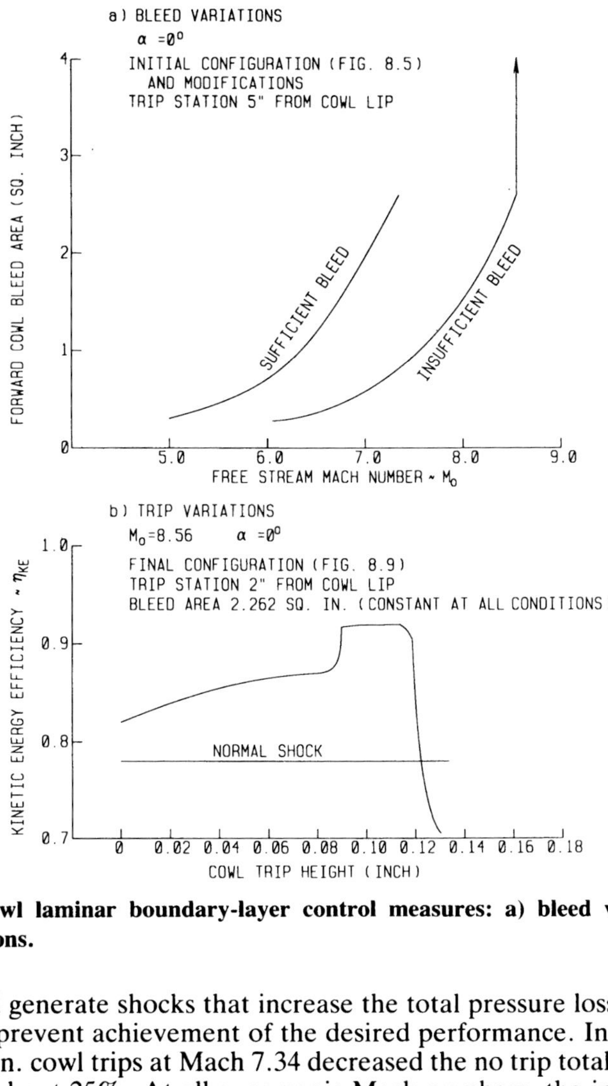

Fig. 8.14 Cowl laminar boundary-layer control measures: a) bleed variations; b) trip variations.

ary layer and generate shocks that increase the total pressure losses in the diffuser and prevent achievement of the desired performance. Installation of the 0.114-in. cowl trips at Mach 7.34 decreased the no trip total pressure recovery by about 25%. At all supersonic Mach numbers, the inlet could not be started with the 0.114-in. cowl trip installed.

Ancillary Data

Wall static pressures measured with taps along both compression surfaces were monitored to determine regions of boundary-layer separation and the effect of bleeds and trips on the boundary-layer behavior. Typical axial static pressure distributions for the final test configuration operating at maximum contraction ratio are plotted in Fig. 8.15 at a low freestream Mach number (upper half) and a high freestream Mach number (lower half). Static tap locations are shown on the sketch at the top of the figure.

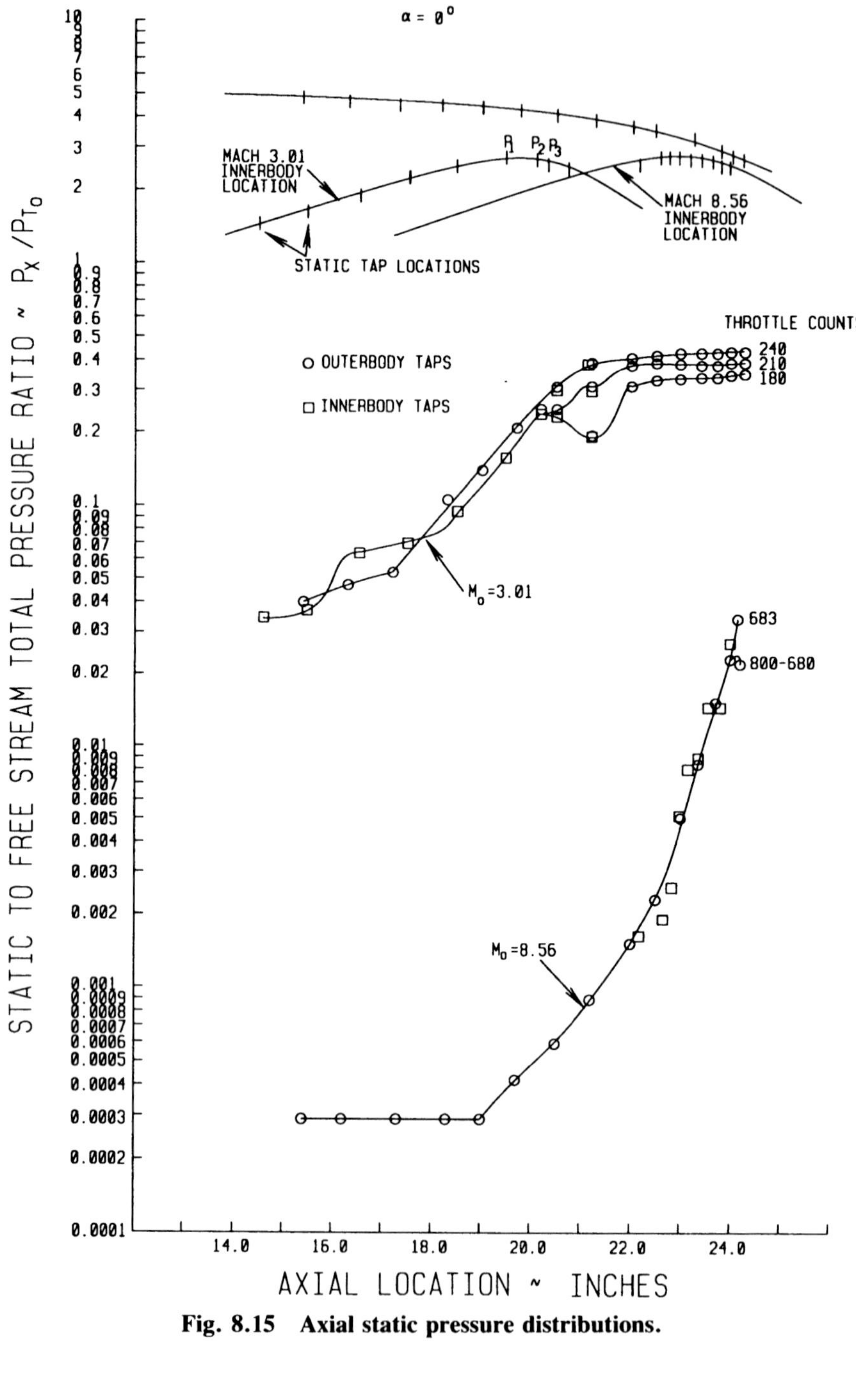

Fig. 8.15 Axial static pressure distributions.

At the lower Mach number, the pressure rise is more gradual and extends over a greater length of the diffuser than at the higher Mach number. The discontinuity in the slope of the inner-body distribution between the second and fifth tap at Mach 3.01 indicates a local separation caused by impingement of the internal cowl lip shock. With forward inner-body bleed the boundary layer reattaches in a relatively short distance. The separation region was more extensive in the original configuration without forward inner-body bleed, but overall total pressure recovery was about the same level in both tests. At Mach 8.56, there is a plateau in the static pressure readings along the forward portion of the cowl. Such a plateau is indicative of a boundary-layer separation that extends well downstream from the trip station and the bleed holes, but the boundary layer attaches to the outer body before the pressure gradient becomes very steep. It is possible that the inner-body boundary layer also separates, but there are no static taps along the forward portion of the spike. Evidently, some separation can be tolerated in the diffuser without sacrificing the desired high level of compression. Figure 8.15 also shows that both compression surfaces contribute the same amount to the process as prescribed in the design procedure.

The effect of throttle position (which simulates changes in the heat release of a flight engine combustion chamber) on the axial static pressure distribution at maximum contraction ratio is included on the figure. The highest throttle count shown represents the position for critical total pressure recovery. As the throttle count is decreased (lower heat release), the static pressures in the subsonic diffuser and throat section decrease in conformance with the change to supercritical operation. The most forward tap affected by the operational change represents the beginning of the actual throat and signifies the dividing line between the regions of the inlet that are responsive to backpressure changes and those that are responsive to contraction ratio changes. Static taps in each region were examined to find suitable sensing parameters for input to automatic controls on backpressure and spike position. A closed-loop spike position control with no errors could achieve maximum contraction at all low Mach numbers between 2.6 and 4.0 with a sensing signal from tap P_1 on the inner body and a reference signal equal to 0.367 Pt^1 generated by an 8-deg two-dimensional isentropic compression surface. At Mach numbers between 4.0 and 6.4, the sensing signal for maximum contraction switches to tap P_2 and a reference signal equal to 0.221 Pt^1 by a 43-deg two-dimensional isentropic compression surface. Above Mach 6.4, P_2 still provides the sensing signal but must be matched with 15.75 Pt^1 generated by a 10.2-deg two-dimensional isentropic expansion. Backpressure control in the supersonic region could be achieved with acceptable resolution by sensing shock position with static tap, P_3. No adequate sensing signal was found in the Mach number range above 6.0, only a limited analysis of control at angle of attack was conducted, and inlet unstart prevention was not studied.

All of the performance data presented heretofore have been the best, or the highest, or the maximum consistent with the maximum achievable contraction area ratio at each freestream Mach number. The final inlet configuration (Fig. 8.9) was also tested at contraction ratios less than the maximum to determine the effect on performance. Inlet total pressure recovery is plotted vs contraction ratio in Fig. 8.16 at each of the test Mach numbers in the two Jet Propulsion Laboratory wind tunnels. The results show that inlet recovery is very sensitive to contraction ratio changes at low Mach numbers and becomes increasingly less sensitive as Mach number increases. The same trend has been observed in other test programs, with two examples[30,32] included in the figure. The one-dimensional flow properties presented in Fig. 8.12 showed that more than 50% of the losses in an inlet operating at hypersonic velocities were chargeable to the hypersonic diffuser. At Mach 8.0, where the hypersonic diffuser total pressure losses are of the order of 80%, the normal shock and subsonic diffuser losses are minor contributors to the overall inlet performance. Decreases in contraction ratio should have only a minor effect on the hypersonic diffuser losses, which continue to dominate the relatively small percentage increases in normal shock and subsonic diffuser losses. As freestream Mach number decreases and the hypersonic diffuser plays an increasingly smaller role, increases in normal shock and subsonic diffuser losses with decreasing contraction ratio become increasingly dominant in establishing overall inlet performance levels. At Mach 3.0, where the central core potential flowfield extends across much of the passage at the throat plane, the throat Mach number and corresponding normal shock recovery are very sensitive to changes in the contraction area ratio.

It was also observed during the test program that the family of inlets could be started in the hypersonic region at contraction ratios well above the theoretical limit as expressed in Eq. (2.16). The results are presented in Fig. 8.17. All of the inlets had open-bleed ports in the hypersonic diffuser.

Critical kinetic energy efficiencies and maximum contraction area ratios for the final inlet configuration at 5-deg angle of attack are plotted against

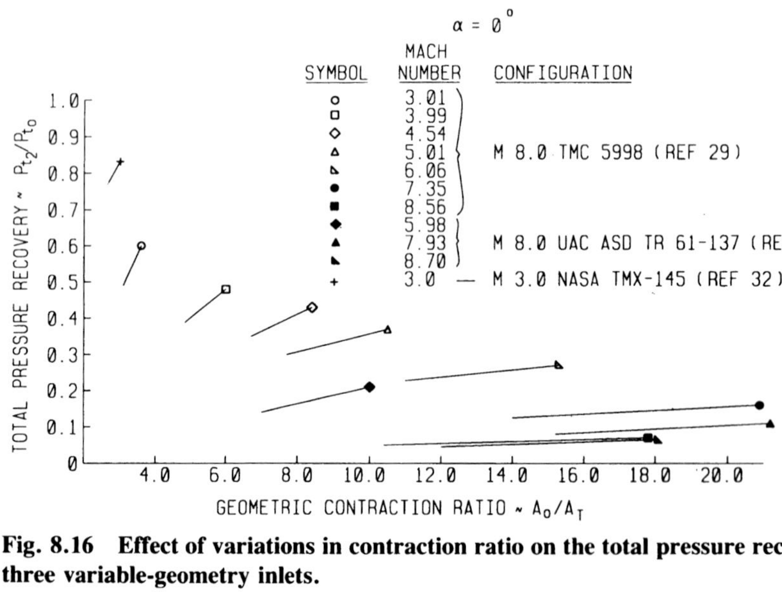

Fig. 8.16 Effect of variations in contraction ratio on the total pressure recovery of three variable-geometry inlets.

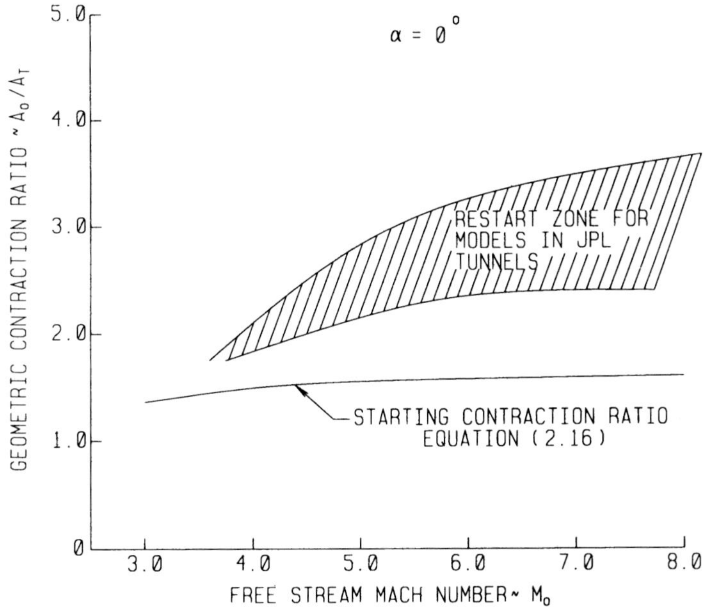

Fig. 8.17 Starting contraction ratios for a family of variable-geometry axisymmetric inlets.

freestream Mach number in Fig. 8.18. For comparison the comparable values at 0-deg angle of attack are reproduced from Fig. 8.10. The 5-deg angle-of-attack n_{kE} data are considerably lower at all Mach numbers but much worse at the lower end, where the contraction ratio agreement is the closest. This apparent paradox seems to offer further proof of the extreme sensitivity of total pressure recovery to the contraction ratio in the lower-Mach-number region. In fact, the sensitivity is increased during angle-of-attack operation, probably because of flow distortions produced by flow asymmetry. The bleed flow rates at 5-deg angle of attack were increased on average more than 50% across the hypersonic Mach number range. Thus, operation of the Mach 8.0 variable-geometry inlet at an angle of attack of 5 deg has the twofold disadvantage of a significant decrease in total pressure recovery and delivered mass flow and a significant increase in external drag contribution.

Flow profiles (not shown) across the subsonic diffuser exit, even with outer-body throat bleeds closed, were relatively undistorted at all hypersonic Mach numbers for critical and supercritical operation at both 0- and 5-deg angles of attack. No flow-straightening devices appear necessary to ensure favorable conditions for fuel injection and combustion.

8.4 Hypersonic-Flight Prospective

All man-made vehicles that have achieved escape velocity have navigated across a sizable portion of the hypersonic regime. Generally, most of the

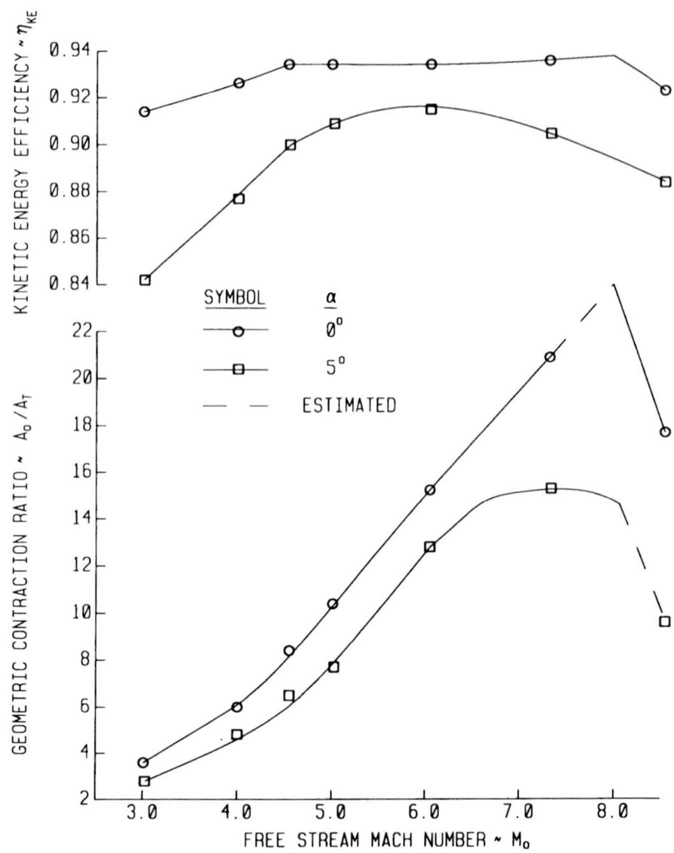

Fig. 8.18 Effect of angle of attack on the overall performance of a Mach 8.0 variable-geometry axisymmetric inlet.

flight has been accomplished in space where small rocket engines are adequate to effect desired velocity changes. From re-entry to landing, the glider-like Shuttle has descended through the hypersonic, supersonic, transonic, and subsonic speed ranges within the Earth's atmosphere; hence, the feasibility of hypersonic operation has been proven, although not under steady-state conditions. Analytical calculations based on ground tests of inlet and combustor/nozzle components have shown that adequate thrust levels with hydrogen-fueled ramjet engines can be generated to accelerate a hypothetical vehicle from low supersonic velocities to sustained cruise at hypersonic Mach numbers.[33] There are still many areas that require extensive exploration before steady-state hypersonic flight within the Earth's atmosphere becomes routine. With the exception of any recent unpublished data which have advanced the state of the art, variable-geometry inlet development status may be succinctly summarized as follows: Inlet tests have been conducted with manually operated models under essentially perfect-gas flow conditions with no capability to simulate realistic flight temperatures or steady-state heat-transfer rates and with boundary-layer control configurations that may be ineffective in a real-gas flight environment.

The spike position control scheme, including the accidental unstart warning and prevent mechanism used in the YF-12 aircraft inlet (Fig. 8.3), could provide a sound basis for automatic control of the hypersonic variable-geometry components with the added complication that different sensing signals would be required over different sections of the Mach number range. Backpressure control by means of normal shock position detection was successfully demonstrated on the BOMARC in flight, and a similar type of control with appropriate sensing parameters could be applied to the hypersonic inlets. More data are needed for inlet angle-of-attack operation and for transient conditions. Also, backpressure response to signals generated by pitot probes mounted on the movable inner body needs to be investigated.

Practical structural materials designed to withstand the pressures and temperatures imposed by hypersonic flight will require some degree of cooling. The fuel tank provides a ready heat sink; thus, fuel circulating through passages in the inlet structure, thence to the fuel injectors, seems to be a solution. Of all ramjet fuels, liquid hydrogen offers the best heat sink, but it is difficult to store and handle. Other ramjet fuels may not have sufficient cooling capability under some flight conditions; hence, alternate or supplemental cooling means may be a necessary complication. Wind-tunnel tests with a water-cooled copper model[30] have shown that cooling can also provide adequate boundary-layer control for achievement of target total pressure recovery levels (Fig. 8.13). Inasmuch as the effects of real gas on the behavior of a boundary layer and, in particular, the separation characteristics are unknown, the efficacy of wall cooling in controlling flight boundary layers is also unknown.

The information given in Ref. 1 shows that real gas effects on the potential flowfield require increased contraction ratio compared with perfect-gas values to achieve a desired throat Mach number. Thus, the surface pressure gradients in a real-gas potential flow environment would be greater than those in a perfect-gas potential flow situation, and boundary layers would be more prone to early separation. As shown in Fig. 8.12, however, viscous effects play a prominent role in the hypersonic compression process such that actual one-dimensional throat Mach numbers are much less than perfect-gas potential flow values, and the pressure gradients imposed in the process do not induce any catastrophic boundary-layer separation tendencies. If viscous effects are equally prominent during compression of a real gas, the additional contraction may not be required, the surface pressure gradients may not be excessive, and the demonstrated boundary-layer control techniques may be applicable, although the amount and location could be different.

Steady-state temperatures in the neighborhood of stagnation points may promote boundary-layer instability and separation which wall cooling could control under most flight conditions. At altitudes above the critical value (Fig. 8.4), wall cooling alone may be insufficient to control the laminar boundary layer, and trips may be a mandatory adjustment to force transition. In other regions of the inlet the cooling flow required for structural integrity may be incompatible with the cooling flow necessary for turbulent boundary-layer control, and other measures, such as bleed, may be essential. Conceptual designs of variable-geometry inlets for hypersonic flight applications should include a combination of cooling, bleed, and trip devices.

Wind-tunnel tests have shown that bleed and wall cooling, applied in separate programs, are both effective in controlling boundary layers, but no combination has been investigated. Test data presented herein imply that there is a best (minimum) amount of boundary-layer control that varies with Mach number and with altitude above critical. Bleed flow rates in excess of the minimum have no significant effect on inlet total pressure recovery but reduce useful mass flow and increase drag. Trip heights on the order of the boundary-layer thickness at the trip station can force the transition from laminar to turbulent flow with little performance penalty, but, at flight conditions where trips are not required, the presence of trips can produce intolerable performance degradation. Successful transition also mandates that the boundary-layer control be applied in the proper location. Axial static pressure distributions (Fig. 8.15) indicate laminar boundary-layer separation along the forward cowl from the lip to a station downstream from the cowl trip and bleed holes at a Mach number of 8.56. Other experimenters[34,35] indicated extensive separation regions in hypersonic diffusers at high contraction ratios and measured total pressure recoveries at the target level. If stable boundary-layer separation could be promoted along most of the compression surface in the diffuser by proper control techniques, the viscous losses could be reduced without sacrifice in total pressure recovery. A major goal of future investigations should be the determination of the least amount of boundary-layer control, in any form, consistent with high performance and the variation of this minimum with Mach number and altitude. Hypersonic inlet models should be designed to incorporate cooling passages that can be opened or closed, bleed ports that can be covered or uncovered, and trips that can be extended or retracted. Another feature for reducing viscous losses at low-Mach-number operation of a hypersonic design is a telescoping centerbody that is collapsed at low Mach numbers and extended as the spike is retracted at high Mach numbers.

Sustained hypersonic flight within the Earth's atmosphere is possible before the end of the 20th century. After boost to a predetermined velocity by some other propulsive device, acceleration and cruise thrust would most likely be supplied by a ramjet engine with a high-Mach-number design, variable-geometry inlet similar to the one described herein. For the best overall performance the inlet should be a fully automatic component with self-contained control functions and actuation mechanisms that can position the spike for maximum contraction at any angle of attack or yaw at any Mach number; change the spike position instantaneously to prevent accidental unstart; change the fuel flow rate to the fuel injectors in response to varying power demands and to prevent subcritical operation; change the fuel flow rate to structural cooling passages as a function of Mach number and altitude; open and close bleed port openings in concert with preselected signals; extend and/or retract trips on the appropriate side of the critical altitude, and extend the telescoping inner body at the proper rate during the acceleration phase. Cost, weight, and reliability may mitigate against such an enormously complicated component. The simplest alternative is an inlet with length and boundary-layer control configuration fixed at the design Mach number with off-design performance penalties that reduce net thrust. However, before any meaningful tradeoffs can be accomplished, the existing data base must be considerably expanded.

9
PROOF OF CAPABILITY

9.1 Methodology

The ultimate proof of capability is flight through the Earth's atmosphere of a structurally sound inlet that delivers the expected performance. However, prior to first flight, it is reasonable and prudent to verify capability with the highest possible degree of confidence using procedures that can be accomplished on the ground. Throughout the text, three procedures have been discussed in connection with the determination of the aerodynamic design and performance of the major components that form a complete inlet.

Analytical procedures involve mathematical solutions to the equations of motion. The Navier-Stokes nonlinear partial differential equations for unsteady, compressible, viscous, heat-conducting fluid in three dimensions cannot be integrated directly; thus, simplifying assumptions and approximations have been used to forge mathematical tools that yield an acceptable accuracy. The inviscid set (Euler) of the equations of motion have been solved for a variety of flow situations to produce one-dimensional channel flow, compressible flow tables and charts, Bernoulli equations, Rankine-Hugoniot shock equations, Laval nozzle, Taylor-Macoll, Prandtl-Meyer expansion, a hodograph method, a method of characteristics, finite difference methods, etc. Simplifying assumptions with viscous flows have led to an accurate prediction of laminar and turbulent boundary layer characteristics, such as thickness, profile, and shear stress.

Empirical methods are solutions based on a body of test data that can be correlated by means of a mathematical expression, a particular curve fit, or even a rule-of-thumb approximation. Generally, the problem involves an analytically intractable flow situation, such as a strong shock/boundary-layer interaction, a potentially high distortion of a flow profile, or an enclosed channel with vortex sheets or slip planes.

Experimental procedures require the measurement of performance parameters during simulation of actual flight with a replica of the real inlet installed in a ground test facility. Problems pertaining to boundary-layer control (transition, separation, ingestion by inlet), angle of attack (yaw), inlet control (i.e., shock positioning), inlet interactions with other com-

Table 9.1 List of inlet development problems and the procedures used in the solution

Procedure →	Analytical	Empirical	Experimental
Flow or design situation	Supersonic diffuser design	Subsonic diffuser design	Boundary-layer control: bleed, wall cooling, trips, effect on throat design
	Forebody/inlet interactions (windward side)	Throat section design	
		Cowl turning radius	
	Capture area		Angle-of-attack (yaw) crossflowing boundary layer
	Pressure recovery Except →	Pressure recovery above design M	Inlet control: shock positioning variable geometry
	Drag		
	Turn & dump recovery at low M_2		Turn & dump recovery at high M_2
	Boundary-layer profiles		Vehicle boundary-layer ingestion by inlet
			Vehicle boundary-layer separation inlet placement
			Adverse inlet/combustor interactions

ponents, and unsteady flow situations (i.e., inlet buzz) are most amenable to acceptable solutions through testing.

Table 9.1 contains a list of the flow and design situations encountered in the development of an inlet and the procedure usually employed (up to the present time) to effect a satisfactory (acceptable) solution that instills a high level of confidence. The three procedures are complementary, and a mix of the three is standard practice during the inlet development cycle. With the advent of more sophisticated computers, direct analytical solutions to more complicated forms of the equations of motion are becoming attainable. Progress in computer capability can shift many of the problems from the empirical and experimental columns to the analytical column on Table 9.1. However, even with a significant shift, many of the problems related to viscosity will remain analytically intractable, and the experimental procedure will continue to be viable.

Both the analytical and empirical procedures employed in the solution of the flow/design problems listed in Table 9.1 have been described in sufficient depth and clarity to furnish a relatively complete set of user-ready tools packaged in the form of equations, figures, tables, and references. Conversely, the experimental procedure has been presented in more abstract terms, such as the following: 1) It is prudent to plan tests with certain hardware variables; 2) the performance can best be determined in tests; 3) the experimental data confirm trends; 4) the correlations are based on data, etc. In order to elevate the experimental procedure description to a level that approaches the user-ready status of the other two procedures, the next two sections of this chapter are devoted to a discussion of specific ground test tools, including the apparatus, objectives, available facilities, and operating protocol.

The last section in the chapter summarizes the traditional roles of the analytical, empirical, and experimental procedures in inlet development; describes the limitations inherent in each procedure; explains the effect of increasing velocity on flow similarity between ground and flight situations; and speculates on the future course of ground-based means of verifying flight capability.

9.2 Test Techniques

There are four types of ground tests for assessing various aspects of the performance capability and operating characteristics of an inlet and verifying design practices:

1) Isolated: an inlet immersed in a freestream flow with a combustor simulator that provides a variation in backpressure to mimic heat-release variations.

2) Installed: an inlet with a combustor simulator mounted on a vehicle immersed in a freestream flow.

3) Semi-freejet: an inlet with a burning combustor and nozzle (complete engine) immersed in a freestream flow.

4) Freejet: complete engine mounted on a vehicle immersed in a freestream flow.

The first two tests are aerodynamic in nature and are usually conducted with scaled pressure models in two types of wind tunnels. Exploratory tests with many hardware variables are customarily conducted in an intermittent-flow wind tunnel, whereas documentation tests of a selected configuration are most compatible with a continuous-flow facility.

The latter two tests are propulsion-oriented and are conducted usually with full-scale hardware in large test cells or propulsion wind tunnels. These engine tests are described in the text only in sufficient detail to complete the inlet development picture between the first exploratory ground test and the first flight test.

An additional type of test for determining contributions of an inlet component to the lift, drag, and stability of a missile is conducted with a force and moment model in a wind tunnel by the external aerodynamics group. Discussions of force models and associated tests are beyond the scope of this text.

The first ground test in an inlet development cycle is generally conducted with an isolated (no vehicle or other engine components in juxtaposition), instrumented model installed in an intermittent-flow wind tunnel. The inlet, consisting of a supersonic diffuser, throat section, and subsonic diffuser, is attached to a calibrated plenum and movable throttle assembly (normally supplied by the facility) that can simulate a variation of heat release from cold flow through stoichiometric, and higher, if necessary. Figure 9.1 portrays a typical isolated inlet model (in this case, the example is a two-dimensional, downward-compressing inlet) and shows the kind and quantity of instrumentation located at different stations throughout the inlet. Aerodynamic contours and internal lines proposed for the flight inlet are replicated in the model, but the structural design of the wind-tunnel article is generally more rugged than the flight prototype to withstand tunnel starting loads and to accommodate the instrumentation and associated tubes, wires, and transducers. There are 37 wall static pressure taps distributed along the inlet surfaces between freestream station 0 and subsonic diffuser exit station 2, where 8 rakes containing 24 total pressure probes are mounted. The total pressure probes are used primarily to determine the amount of flow distortion that would be delivered to the engine com-

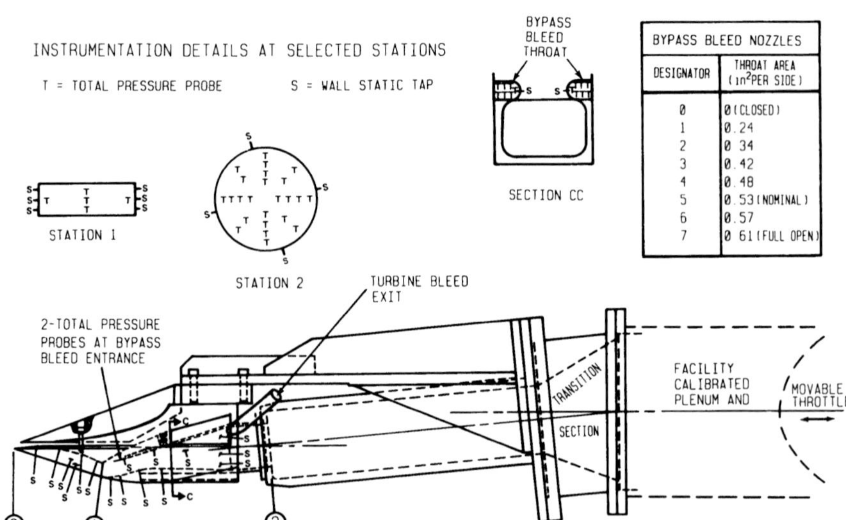

Fig. 9.1 Typical isolated inlet model showing distribution of instrumentation.

bustion chamber over a wide range of operating conditions. The static taps can be used to trace the movement of the terminal normal shock as the backpressure varies with throttle position and to show static pressure plateaus commensurate with boundary-layer separation. Mass flow through the plenum is known for every throttle position, and mass flow through any bleed device can be measured with appropriately located instrumentation. The inlet shown in Fig. 9.1 features a bypass bleed that consists of a flush slot bleed entrance located on the main compression surface immediately downstream from station 1, a bleed plenum or collector, and two aft-facing nozzles with throats at section CC. There are 9 wall static taps along the internal bleed passages from the entrance to the nozzle throat, 2 total pressure probes immediately downstream from the entrance, and 14 total pressure probes in the exit nozzle throat plane. In addition, the turbine bleed (the exit shown in the figure) has two statics and two totals to assist in measuring the performance of this flow extraction device.

Some of the hardware variables that would be suitable for investigation with this particular inlet configuration include different size exit nozzle plugs for the bypass bleed, and a total of seven have been constructed to give the eight bleed flow conditions listed in the table in Fig. 9.1. Flush bleed holes across the main compression ramp and along each side plate upstream from station 1 (not shown in the figure) could be closed systematically by means of dental plaster to vary supersonic boundary-layer control. Primary test objectives would be the validation of design techniques and performance estimates over the desired range of test conditions and the determination of the effects of various bleed configurations on inlet performance. A primary post-test objective would be the selection of a "best" configuration for a subsequent documentation test.

For missiles with closely integrated inlets, the documentation test involves an instrumented pressure model inlet configuration installed on a vehicle model with faithful replication of forebody geometry, wings, or other protuberances that can influence the airflow captured by the inlet. Most commonly, the installed model is designed with removable forebody, wings, and other protuberances, so that the isolated and installed inlet tests can be conducted in one wind-tunnel entry with one relatively simple change to accomplish the conversion. Thus, the same scale relative to the flight article is preserved in the model hardware for both kinds of tests. Determination of maximum model size is discussed in the next section.

The two-dimensional downward-compressing inlet of Fig. 9.1 is shown installed on a representative vehicle in Fig. 9.2. In addition to the instrumentation shown previously on the isolated model, there are 14 wall static taps (7 on each side) on the boundary-layer diverter wedge and 8 taps (4 on each side) along the external surfaces of the bypass bleed exit nozzle. Readings of the static pressure taps along the boundary-layer diverter can be used in the calculation of diverter drag as discussed in Chapter 7. A primary hardware variable in this type of model could be vehicle boundary-layer diverter height. The primary test objectives would be the determination of vehicle/inlet interface over a range of test conditions including angles of attack (or yaw), the verification of flowfield code predictions,

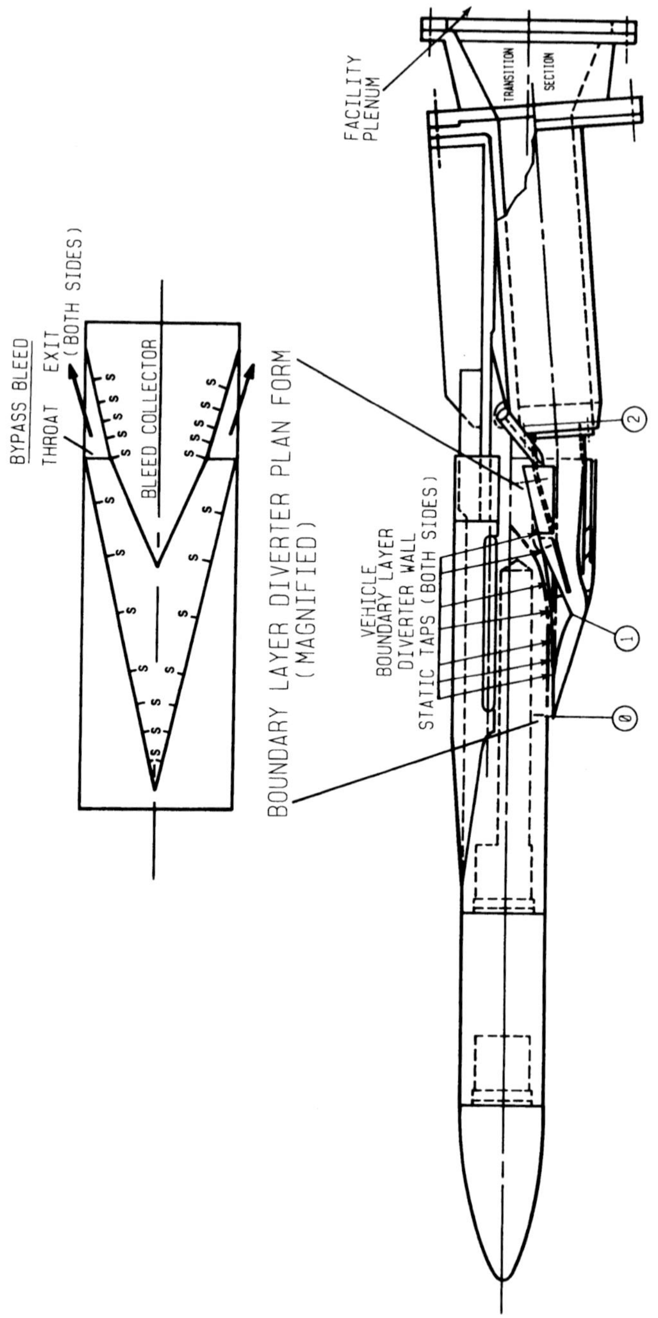

Fig. 9.2 Typical inlet model installed on a representative vehicle model.

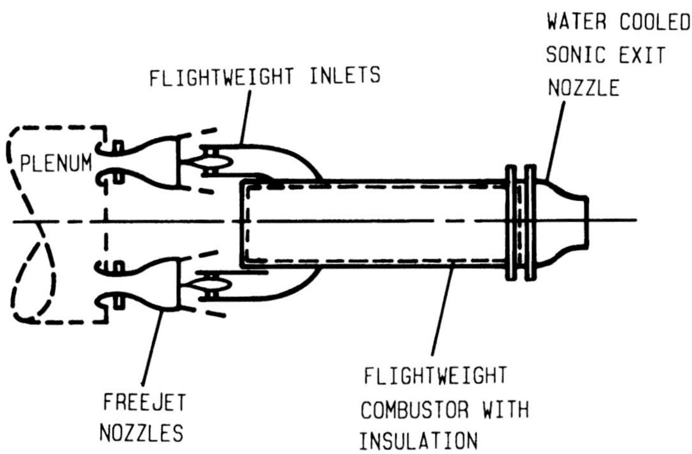

Fig. 9.3 Typical semi-freejet test setup.

and the measurement of performance variation with various amounts of vehicle boundary-layer ingestion by the inlet.

In a typical engine development program, the inlet component, the combustor/nozzle components, and, for liquid-fueled engines, the fuel management subsystem are investigated simultaneously along parallel paths. In order to determine interactions between the major components, the best configurations selected from the independent investigations are mated for the propulsion test in either a semi-freejet setup, as shown in Fig. 9.3, or freejet setups, as shown in Figs. 9.4 and 9.5, with the engine installed on a vehicle in a large test cell and a propulsion wind tunnel, respectively.

The engine illustrated in Fig. 9.3 consists of dual axisymmetric inlets dumping the flow through ports in the sides of a flight weight insulated combustor and a water-cooled sonic exit nozzle. The supersonic diffusers of the two inlets are completely immersed in the rhombuses of the two

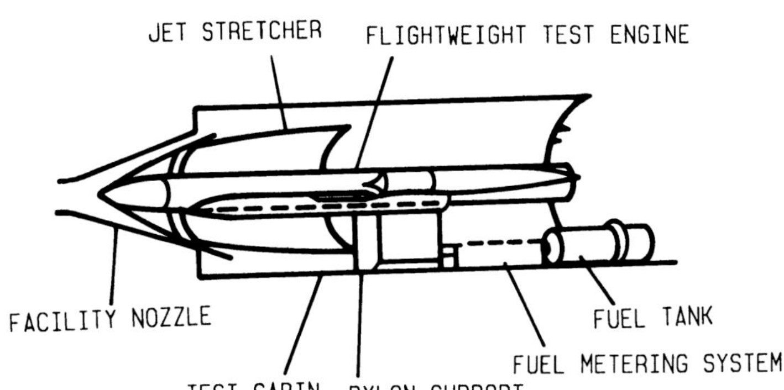

Fig. 9.4 Installed engine with aft inlets mounted in large freejet cell.

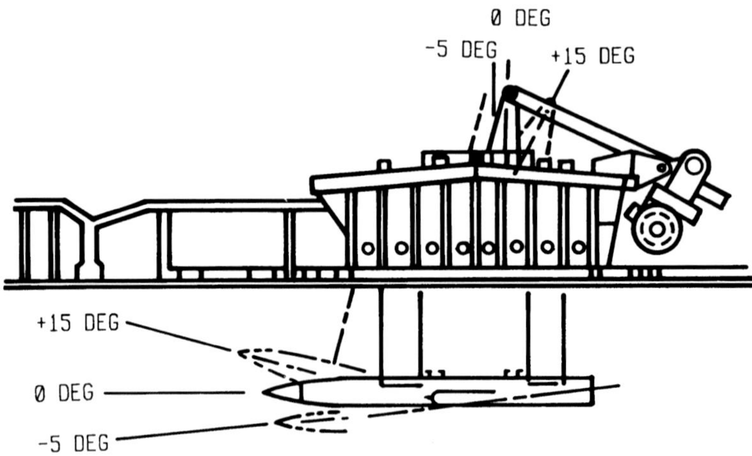

Fig. 9.5 Installed engine with aft inlets mounted in propulsion wind tunnel.

facility nozzles which could be designed at any one of a number of supersonic Mach numbers. Engine/vehicle interactions cannot be determined in this setup. Instrumentation is nonintrusive and consists primarily of three or four static pressure taps at each of several strategic stations in the engine. The test objectives are the determination of inlet/combustor interface and the documentation of engine performance over a range of heat-release values at a number of simulated flight conditions. Another objective pertinent to successful engine operation includes the examination of structural durability.

Both freejet setups shown in Figs. 9.4 and 9.5 feature an engine with dual axisymmetric inlets mounted on each side of the vehicle near the midbody station and dumping the flow into a combustor (not shown) within the vehicle body. In the test cell installation (Fig. 9.4), the aft-mounted inlets require a jet stretcher to lengthen the rhombus of the facility nozzle, but the jet stretcher inhibits angle-of-attack operation over the desired range. The wind tunnel with a larger test rhombus (as shown in Fig. 9.5) permits angle-of-attack operation with the aft inlet installation.

Because of size, shape, number, or location on the vehicle, many inlet configurations are not amenable to the semi-freejet test mode. For example, a chin inlet, where the vehicle forebody is an integral part of the inlet compression surface, is most compatible with the freejet mode, as illustrated in Fig. 9.6. This inlet installation requires no jet stretcher; thus, angle-of-attack performance can be determined in a large test cell without recourse to the propulsion wind tunnel. Test objectives of the freejet mode are the determination of installed engine performance over a range of heat-release values over the widest possible range of vehicle angles of attack at a number of simulated flight conditions. Other objectives include assessment of durability, thermal protection, and, for integral rocket/ramjet engines, demonstration of transition.

9.3 Test Facilities

An inventory of the major aeronautical ground research facilities available in the United States in 1971 is contained in Ref. 36, and four categories are listed: 1) wind-tunnel facilities, 2) air-breathing engine test facilities, 3) structural and environmental facilities, and 4) engineering flight simulation facilities. An updated survey[37] lists the major facilities available in the United States, Canada, Japan, and several European countries in January 1985, and the catalog is composed of two volumes: 1) wind-tunnel facilities and 2) air-breathing propulsion and flight simulator facilities. Many facilities listed in Ref. 36 do not appear in Ref. 37, and it is probable that such facilities have been permanently deactivated. In both references wind tunnels are subdivided in accordance with operating Mach number range into subsonic ($M \sim 0$–0.99), transonic ($M \sim 0.5$–1.4), supersonic ($M \sim 1.4$–5.0), and hypersonic ($M > 5.0$). Only the last two subdivisions will be discussed in any detail in this chapter. Generally, the descriptive data provided in the references is limited to supersonic tunnels with minimum test section size of 2 ft × 2 ft and hypersonic tunnels with a minimum test section size of 1 ft × 1 ft. No combustion is permitted in any of the wind tunnels, and up to about Mach 5.0 the supply air is ambient temperature. At and above Mach 5.0, the supply air is heated to avoid liquefaction and condensation shocks. Wind tunnels of both the closed-circuit continuous-flow type and the intermittent blowdown to atmosphere type are listed and described.

The second facility category of interest in the development of an inlet includes freejet altitude cells and propulsion wind tunnels where test engine combustion is permitted, and the supply air is heated to simulate the total temperature that will be encountered in flight. A range of Mach numbers can be achieved usually with a series of interchangeable freejet nozzles, and a range of flight conditions can be simulated by regulating the supply pressure and temperature and evacuating the test chamber.

During the course of supersonic integrated inlet development by means of the experimental procedure, as many as five different ground test facilities can contribute to the desired proof of capability. An experimental development path and five representative test facilities with salient operating characteristics are presented in Table 9.2. The path shown in the table is illustrative only, and it is neither essential nor recommended that all inlet development programs follow this path. Furthermore, as noted in Ref. 37, many other facilities may be substituted for each of the ones shown in the table.

Exploratory tests of a new inlet configuration in both the isolated and installed modes are conducted in the LTV (Ling Temco Vought) High Speed Wind Tunnel (HSWT), which is a blowdown type with a 4 ft^2 test section. The maximum model size permitted in any tunnel is a function of the test section area blockage during tunnel start at 0-deg model angle of attack, interference between model shocks and tunnel wall boundary layers at maximum model angle at attack as determined by flight maneuver requirements, and model station at which reflected shocks impinge on the

Table 9.2 An experimental development path for supersonic integrated inlets and representative ground test facilities with salient operating characters

Test	Facility	Test section	Mach no.	Reynolds no. ($\times 10^6$/ft)	Pt_0, psia	q, psf	Tt_0, °R
1. Isolated & installed inlet exploratory development	LTV HSWT intermittent blowdown to atmosphere	4 ft × 4 ft × 5 ft Long	0.2–5.0	2–38	19.8–367.5	150–5000	540–620
2. Isolated & installed inlet performance documentation	NASA Langley UPWT closed-circuit continuous flow	Two legs 4 ft × 4 ft × 7 ft Long	1.47–2.86 2.29–4.63	0.5–12.2 0.5–9.5	3–56.9 3–150	90–2670 90–1710	560–760 560–760
3. Isolated engine inlet/combustor/ nozzle interactions	MJL full-scale freejet altitude cell 8 run time function flow rate	14 ft diam × 80 ft Long	0.8–7.6 Nozzle diam 0.58–2.67 ft		215		5000 Vitiate 920 Clean
4. Installed engine vehicle/engine interactions	AEDC APTU intermittent blowdown	16 ft diam × 42 ft Long	0–4.5 Nozzle diam 2–3.5 ft		300	400–10,000	2000 Vitiate
5. Complete missile maneuvering interactions	NASA LEWIS wind tunnel continuous open-circuit on propulsion cycle	10 ft × 10 ft × 40 ft Long	2–3.5	2.1–2.7	1.4–34.7	500–600	520–1150

model. For tests of an inlet installed on an air-launched missile, such as that shown in Fig. 7.1, for example, the model scale in the 4-ft tunnels will be on the order of one-fourth to two-thirds. These model scales will allow up to 25-deg angle of attack with installed inlet models. Much larger scales are possible with the isolated inlet models, but this arrangement requires two different models for isolated and installed tests with two separate tunnel entries at increased time and cost.

In a typical tunnel installation the model is bolted to the facility supplied calibrated plenum/throttle assembly, as shown in Fig. 9.1, and the instrumentation lines are connected to the facility data acquisition system through strain-gage transducers (up to 110 pressure channels can be connected in the HSWT). The tunnel is started with the model at 0-deg angle of attack and the throttle plug open to simulate a cold flow engine situation. The throttle plug is advanced automatically through a series of 12–14 preset positions that span inlet operation from cold flow supercritical through critical to an unstable subcritical condition. The throttle is automatically retracted as the model automatically rotates to the next preset angle-of-attack position, and the throttle plug automatically advances again through the preset positions. Usually, about 12–14 throttle plug positions at each of five model angles of attack can be explored in about a 40-s blow, and at each of the 50 or 60 points, all transducers can be scanned about three times. During facility pumpup, which is of the order of 1 min for each second of blowdown, the tunnel nozzle blocks can be adjusted for a new Mach number, and the model configuration can be changed. Flexible model mechanical design permits a multitude of rapid changes with multiple hardware variables, including forebody removed or installed, bleed holes opened or closed, bleed exit area changed, cowl lip angle changed, and boundary-layer diverter raised or lowered. The entire process can be repeated at every freestream Mach number at which the tunnel has been calibrated.

After the exploratory isolated and installed inlet tests are completed, the same model can be installed in the NASA Langley Unitary Plan Wind Tunnel to complete performance documentation test with the best of the hardware variables identified in the HSWT tests. The continuous-flow wind tunnel permits a much more precise definition of the inlet critical operating points at each selected angle of attack at each Mach number. This type of tunnel also facilitates the investigation of control features, such as shock position and backpressure, and, for a variable-geometry inlet, determination of the operating contraction ratio. The model is installed in the same manner as described earlier, and the data acquisition system consists of 100 analog and 40 digital channels. Throttle position and angle-of-attack variation can be manually controlled, and a more complete map of inlet performance can be obtained in one run at each freestream Mach number.

At the point in the development path where the inlet is integrated with the combustor/nozzle components, full-scale, and very often, flight weight engines are tested over a range of altitudes and Mach numbers expected to be encompassed in a flight trajectory and over a range of fuel-to-air ratios commensurate with thrust levels from cruise to acceleration. Prior to any hot test, the engine mass flow characteristics can be determined with a calibrated plenum/throttle arrangement similar to that used in the inlet wind-tunnel tests. In particular, the average of three or four static pressure tap reading at each of several key stations in the engine as a function of changes in backpressure is necessary in order to calculate total pressure ratios and deduce total temperature levels. During initial cold flow tests, total pressure rakes may be inserted at station 2 to verify the flow profile shapes measured in the scaled inlet model tests. With liquid-fueled engines hot tests are conducted at several fuel flow rates between the lean blowout limit and the stoichiometric value at each simulated trajectory condition. In Table 9.2 the isolated engine test is shown as in entry in the Marquardt Jet Laboratory (MJL) Cell 8, which is a very large facility capable of broad Mach number and altitude simulation. Full-scale installed engine (complete missile) tests could also be conducted in cell 8. For the purpose of emphasizing facility interchangeability, an installed engine is shown as an entry in the Arnold Engineering Development Center (AEDC) Aerodynamic and Propulsion Test Unit (APTU). Figure 9.6 depicts a schematic of an engine with a chin inlet installed on a full-scale vehicle in the APTU. For the chin inlet installation, a meaningful angle-of-attack performance can be obtained. However, with aft-mounted inlets the facility nozzle rhombus must be lengthened to encompass inlet as well as vehicle forebody, and a jet stretcher such as that shown in the sketch in Fig. 9.4 is used. With the jet stretcher in place, accurate angle-of-attack performance measurement is impossible. Thus, a propulsion wind tunnel with a larger test rhombus, such as the NASA Lewis 10 ft $\times$ 10 ft wind tunnel, is a necessary adjunct for missile angle-of-attack performance measurement with aft inlets.

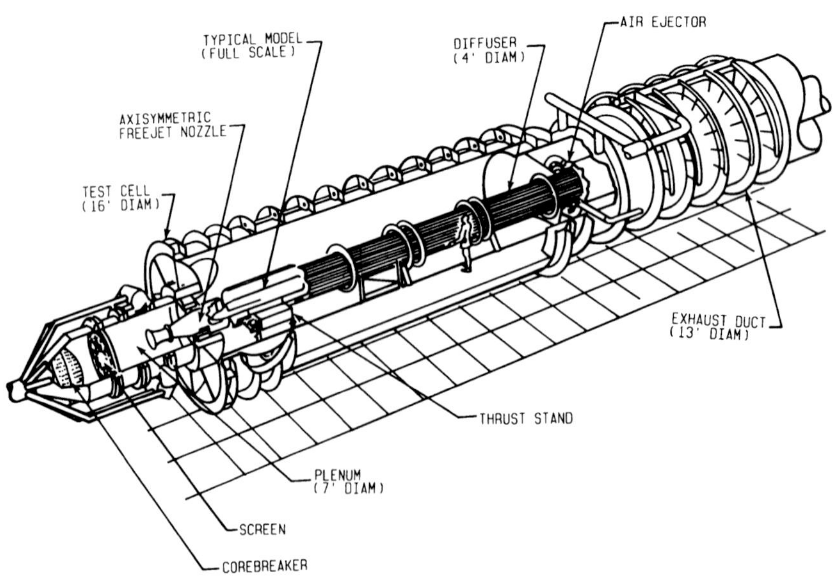

Fig. 9.6 Installed engine with chin inlet mounted in the AEDC Aerodynamic and Propulsion Test Unit (APTU).

The development path for inlets designed to operate in the hypersonic regime has been limited, at least through the 1970s, to isolated testing. Some of the hardware variables, instrumentations, and test techniques are described in Chapter 8. A selected cross section of ground test facilities currently operational in the United States for hypersonic inlet testing has been extracted from Ref. 37 and shown in Table 9.3 in three groups: continuous, intermittent, and shock. There is a fourth group of facilities, which use working fluids other than air (e.g., methane, helium, nitrogen, and Freon), but these are oriented toward the investigation of re-entry conditions at very high Mach numbers (above 15).

The continuous-flow tunnels generally have larger test sections; but offer less Mach number range than the intermittent-flow tunnels. All of the tunnels in both categories are limited in supply pressure (maximum Pt_0 = 3,000 psia) and temperature (maximum Tt_0 = 2250°R). The shock tunnels have obvious advantages in size, Mach number range, and supply pressure and temperature, but these offer only milliseconds of run time. Shock tunnel run times, typically of the order of 5 ms, are too brief to complete the formation of subsonic flow elements, and, in particular, the boundary layer, so that flow separation problems involving low-velocity circulation of significant masses of air are not discerned.

The JPL 21-in. hypersonic continuous-flow tunnel, which was described in Chapter 8, is scheduled to be moved to NASA Lewis Lab and reactivated in the early 1990s. The MJL Cell 8 shown in Table 9.2 has hypersonic capability and is suitable for inlet testing.

9.4 Evaluation

The traditional method for developing any inlet involves a series of tasks that follow a logical progression. First, currently available analytical and empirical tools are used to define aerodynamic contours calculated to satisfy performance requirements over a range of flight operating conditions. Concurrently, estimated performance numbers are generated at each operating condition to provide goals during the development process. Second, a model inlet is fabricated in the largest size that can be installed in the ground facilities selected to simulate as closely as possible the prescribed range of flight operating conditions. Third, the inlet model is tested, and performance is measured at each simulated operating condition. Finally, the measured performance is compared with estimated performance goals. If all procedures are properly applied, there should be acceptable (within the limits of experimental accuracy) correlation between the two sets of numbers, and, on average, these should be a reasonable approximation of performance to be expected in flight. If an exact solution to inlet performance is defined as the performance determined in actual flight, the closest approach in ground-based pursuit of the solution should be achieved at the low subsonic end of the speed regime. As speed increases, ground-based solutions should become increasingly inexact. The problem is portrayed in Table 9.4, where variations in some significant aspects of inlet hardware, flow properties, and wind tunnels are given as a function of

Table 9.3 Selected cross section of hypersonic wind tunnels operational in the United States in 1988

	Test section	Mach no.	Reynolds no. ($\times 10^6$/ft)	Pt_0, psia	q, psf	Tt_0, °R
Continuous flow						
1. AEDC A	40 in. × 40 in.	1.5–6[a]	0.3–9.2	1.5–200	50–1800	530–750
2. AEDC B	50 in. diam	6 or 8	0.3–4.7	20–850	49–600	700–1350
3. AEDC C	50 in. diam	10	0.3–4.7	2000	40–475	1650–2250
4. NASA 1251 A	31 in. × 31 in.	10	0.4–2.4	250–1800	60–400	1810
Intermittent flow						
1. NASA LRC	20 in. × 20.5 in.	6	0.5–10.5	30–550	69–1264	810–1018
2. NASA LRC	18 in.	8	0.1–12	15–3000	100–2643	1160–1510
3. WAL, WPAFB	20 in. diam	12, 14	0.4–1.0	800–1600	40–120	1800–2000
4. NSWC #8	22.5 ft diam	5–8	100–7000	73.5–2205	0.7–60	600–1460
Shock						
1. CALSPAN 96″	24–96 in. variable	6.5–24	0.001–75	1000–20,000	20,000	1300–11,500
2. CALSPAN 48″	24–48 in. variable	5.5–20	0.004–50	200–5400	5700	1100–5800

[a]Listed as supersonic, but has hypersonic capability.

freestream Mach number divided into subsonic, transonic, supersonic, and hypersonic regimes. At Mach numbers below about 0.5, in the domain of the piston engine driving a propeller, the inlet is a simple scoop that guides air to the carburetor, the flow may be treated as an incompressible, perfect fluid, and wind tunnels are available to accommodate full-scale engines and propellers, and mountings thereof.

However, with increasing Mach number inlet shapes become more complex, and, as discussed in previous chapters, throat areas become smaller, boundary layers become thicker, shock/boundary-layer interactions become stronger, and all aspects of boundary-layer control become more difficult. Also, as Mach number increases, the flow equations must be increasingly more complex to accommodate the changing inlet shapes as well as compressibility, shocks, real-gas flow, and imperfect-gas effects.

Heat transfer can be accounted for analytically, but viscosity remains an enigma. Wherever the viscous flow domain becomes significantly large with respect to the neighboring potential flow domain (e.g., a compression surface boundary layer with the same thickness δ as the inlet throat height h or a separated boundary layer in the subsonic diffuser), the orderly flow process is disrupted, and analytical codes are useless. Furthermore, no analytical procedures exist to predict the requirement for boundary-layer control or the best location or amount of control.

If the equations for unsteady, laminar flow of viscous, compressible, heat-conducting air around an arbitrary body are nondimensionalized, a set of similarity parameters can be determined. In order to have complete steady-state similarity between two flow situations (i.e., a flight inlet and its corresponding wind tunnel model), five parameters (Mach number, Reynolds number, Prandtl number, and two boundary conditions for heat transfer) must be satisfied simultaneously. A sixth parameter involving time must be included for similarity of flows during transient operations. All six parameters are functions of *temperature*; Reynolds and Mach number parameters are also functions of *pressure*; Reynolds number, one heat-transfer boundary condition, and the time parameters are dependent on length ratio or *model scale*; and the two boundary conditions and time parameters are affected by body *material*. Exact similarity is only possible with a full-scale model, and the error in flow similarity becomes progressively larger as model scale diminishes. Wind-tunnel *temperature* and *pressure* and model *scale* and *material* are the four controllable variables in ground testing.

Because the power to drive a wind tunnel or recharge storage tanks in a reasonable time increases with increasing Mach number, there is a practical limit on wind-tunnel test section size, and the size limit decreases in the manner shown in Table 9.4. It is likely that full-scale hardware can be accommodated in the low subsonic wind tunnels, and the odds are increasingly against full-scale models as Mach number increases. In the Mach number range of 2.0–4.0, it was previously estimated that model scales for inlets installed on air-launched missiles would be on the order of $\frac{1}{4}$ to $\frac{2}{3}$. Thus, exact similarity for the complete set of parameters is impossible, but the Mach number condition can be satisfied exactly, and the Prandtl

Table 9.4 Variation of some significant parameters with Mach number

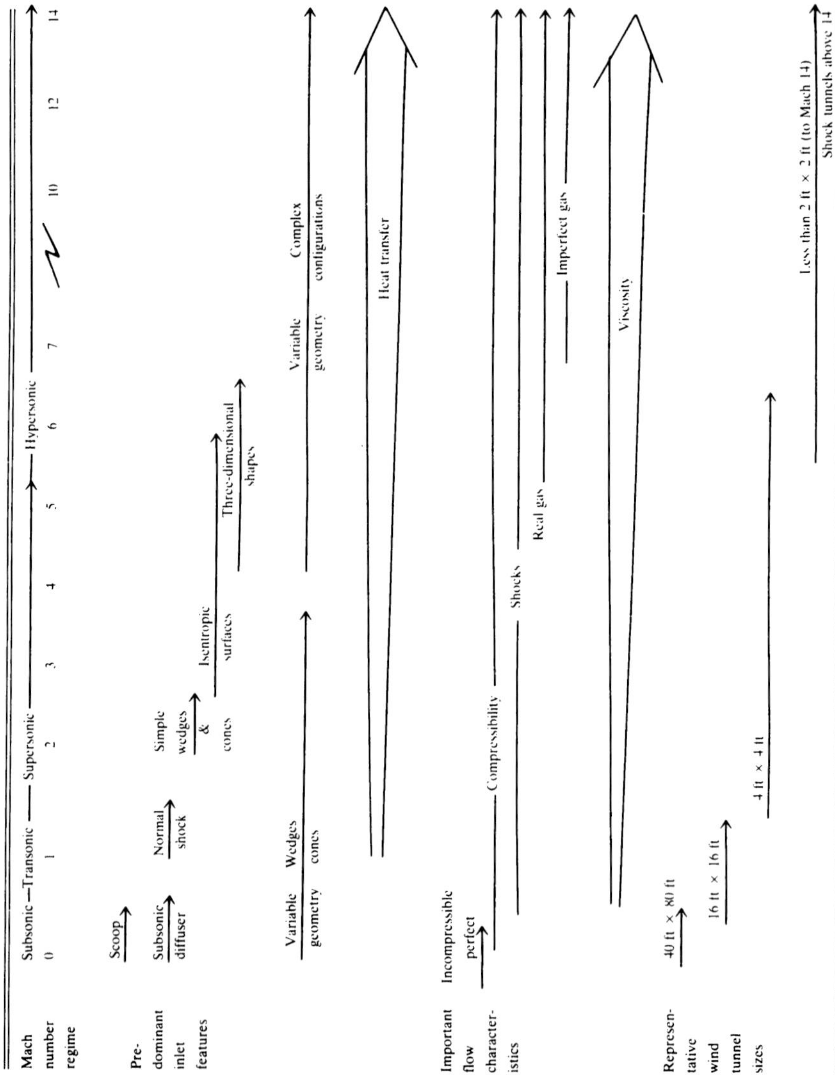

number variation between tunnel and flight will be within the limits of experimental accuracy. Although the range of Reynolds numbers achievable in the wind tunnel will be less than that in flight, flow similarity will be reasonable as long as the tunnel Reynolds numbers exceed the values defined in Fig. 8.4. The two heat-transfer boundary condition parameters will be increasingly in error with increasing Mach number, but flight tests in the range of Mach 2.0–4.0 have not revealed any serious performance deficiencies that could be traced to heat-transfer problems.

In the hypersonic regime, where wind tunnels are limited in test section size, supply pressure, and temperature (Table 9.3), the conditions for simultaneous similarity of all the parameters within a desirable error margin are unattainable. Mach number similarity can be exact at all Mach numbers at which the ground facilities have been calibrated. With increasing Mach number, Reynolds numbers achievable in the tunnels will diverge further from flight values, so that the models will be operating at simulated altitudes above the critical value for laminar boundary-layer separation (Fig. 8.4). It is conceivable that boundary-layer control will be required on the model but not on the flight inlet. However, if proper boundary-layer control procedures are applied, the performance measured in the tunnel can be valid. Prandtl number decreases slowly with increasing temperature (12% change between 0 and 1600°F) so that substantial differences between tunnel and flight temperatures should not produce large errors in similarity for the Prandtl number parameter. As the Mach number increases across the hypersonic region, there will be essentially no similarity in the boundary conditions for heat transfer, and real-gas effects will not be duplicated.

Perhaps one day the key to a direct solution of the Navier-Stokes equations will be discovered, and viscous effects will be analytically predictable. However, until that day the wind tunnel will continue to be an essential tool in the inlet development process. An increasing body of test data should provide rigorous proof of many empirical procedures, which will become rules or laws and be incorporated into the analytical set. The analytical procedures will continue to be essential for defining contours and establishing performance goals, but the experimental procedures will be necessary for determining the detailed methods for attaining the goals. In the Mach number range of 0–14, it is possible to simulate steady-state flight conditions with varying degrees of exactitude with a wide variety of ground test facilities. The Mach number can be exactly simulated at all facility calibrated flow conditions. The error in Prandtl number will be negligible in the subsonic region and gradually increase with increasing Mach number, but even at Mach 14 should not negate satisfactory flow similarity. Reynolds number similarity can be exact across the subsonic and transonic regions, but the maximum Reynolds numbers available in supersonic and hypersonic wind tunnels will most often simulate altitudes higher than flight values. Regardless of simulated altitude, if boundary layers remain attached to compression surfaces, flow similarity will be preserved. The boundary conditions for heat-transfer similarity can also be exact in the subsonic and transonic regimes, but tunnel values will diverge further and further from flight values as speed increases. Neither heat-transfer nor real-gas effects can be reasonably simulated across the high supersonic and hypersonic regimes. A more exact simulation could possibly be achieved in wind tunnels with a working medium other than air and with models fabricated from a different material than the flight prototype. It is speculated that structural cooling requirements can mitigate any adverse heat-transfer effects; hence, even significant errors in similarity of the two boundary conditions for heat transfer will not invalidate performance results achieved with uncooled inlet models. Furthermore, as discussed in Chapter 8, if viscous effects predominate in real-gas flow as in perfect-gas flow, inlet model performance measured on the ground can be a reasonable approximation of flight performance.

It is doubtful that transient conditions associated with acceleration can be simulated in real time in ground test facilities, or that the dynamics of wind gusts or wind shear can be approximated accurately. Also, for variable-geometry inlets, wind-tunnel models will probably have different feedback characteristics for unstart prevention mechanisms and different responses to accidental unstart than flight inlets. Obviously, some solutions can only be determined in flight, but even with the similarity deficiencies noted earlier, ground testing can provide a relatively accurate indication of flight capability over most of the operating envelope.

Isolated and installed inlet tests will continue to represent the heart of the experimental programs in inlet development. For interactions with the combustor/nozzle, including any combustor-induced pressure oscillation impact on inlet performance, the simplest possible test setup (either semi-freejet or freejet consistent with the configurational form and constraints) would seem adequate.

REFERENCES

[1]Ames Research Staff, "Equations, Tables and Charts for Compressible Flow," NACA Rept. 1135, 1953.

[2]Kennedy, E. C., "New Mach Number Tables for Ram-Jet Flow Analysis γ = 7/5 and γ = 9/7," Convair Ordnance Aerophysics Lab., OAL Memo 50–1, Aug. 1955.

[3]Paladin, J. C., "Results of Aerodynamic Burner Drag Tests Conducted at USCEC between March 17 and May 2, 1958," Marquardt Aircraft Co. Rept. M-1754, Feb. 1959.

[4]Liepman, H. W., and Puckett, A. E., *Introduction to Aerodynamics of a Compressible Fluid*, Wiley, New York, 1947 (Galcit Aeronautical Series).

[5]Dailey, C. L., and Wood, F. C., *Computation Curves for Compressible Fluid Problems*, Wiley, New York, 1949.

[6]Kopal, Z., "Tables of Supersonic Flow Around Cones of Large Yaw," Massachusetts Institute of Technology, T R-5, Contracts Ord. 8555 and 9169, 1949.

[7]Ferri, A., *Elements of Aerodynamics of Supersonic Flows,* Macmillan, New York, 1949.

[8]Connors, J. F., and Meyers, R. C., "Design Criteria for Axisymmetric and Two-Dimensional Supersonic Inlets and Exits," NACA TN-3589, Jan. 1956.

[9]Eberly, P. G., "Summary of Data from Tests of the AM45 One-Seventh Scale Aerodynamic Model of a High Compression Inlet for the X RJ 59 Ramjet at MJL During December 1955," Marquardt Co., Rept. M-1567, Dec. 1956.

[10]Luidens, R. M., and Flaherty, R. J., "Use of Shock-Trap Bleed to Improve Pressure Recovery of Fixed - and Variable - Capture Area Internal Contraction Inlets, Mach 2.0 to 3.0," NACA RM E58D24, Aug. 1958.

[11]Stitt, L. E., and Obery, L. J., "Performance of an All-Internal Conical Compression Inlet with Annular Throat Bleed at Mach Number 5.0," NACA RM E58E14, Aug. 1958.

[12]Scherrer, R., Lundell, J. H., and Anderson, L. A., "Viscous Flows in Inlets," NACA A58D17a, June 1958.

[13]Bendot, J. G., Heins, A. E., and Welch, K. E., Supplement I: "AM 115 E and F (Series II) and G Inlet Test Results," Feb. 1965; Supplement II: "Technical Description of an Analytical Method of Determining Shapes of Inlet Elliptical and Near Elliptical Capture Areas," Supplement II, Appendix III: "Design Procedures and Digital Machine Programs," ASD TR-64–91, Vol. 21 of Technical Summary Rept. for 1964 on a Low Altitude Propulsion System.

[14]McLafferty, G. H., Krasnoff, E. L., Ranard, E. D., Rose, W. G., and Vergara, R. D., "Investigation of Turbojet Inlet Design Parameters," United Aircraft Corporation Rept. R-0790–13, Dec. 1955.

[15]Thomas, A. N., Jr., "Inlets," *Tactical Missile Aerodynamics*, Vol. 40, edited by M. J. Hemsch and J. N. Nielsen, Progress in Astronautics and Aeronautics, AIAA, Washington, DC, 1986, pp. 129–167.

[16]Clauser, F. H., "Turbulent Boundary Layers in Adverse Pressure Gradients," Journal of the Aeronautical Sciences, Vol. 21, No. 2, Feb. 1954, pp. 91–108.

[17]Kuehn, D. M., "Experimental Investigation of the Pressure Rise Required for Incipient Separation of Turbulent Boundary Layers in Two-Dimensional Flow," NASA Memo 1–21–59A, Feb. 1959.

[18]Connors, J. F., and Anderson, L. A., Jr., "A Two-Dimensional External-Internal Compression Inlet with Throat Bypass at Mach 3.05," NASA Memo 10–3–58E, Dec. 1958.

[19]Boerem, R. T., Harkins, W. G., Kirby, C. C., and Rosevear J. E., "Critical Components Demonstration Tests for a Chemical Low Altitude Ramjet Engine," The Marquardt Co. Rept. 6088, Feb. 1965.

[20]Brown, J. E., "Results of Series II Test of the AM52 Translating Cowl Inlet Model at BRL, Aberdeen Proving Grounds During February and March 1957," The Marquardt Co. Rept. M-1633, Nov. 1957.

[21]Hoerner, S. F., *Aerodynamic Drag*, published by the author, 1951.

[22]Rubesin, M. W., Mayden, R. C., and Varga, S. A., "An Analytical and Experimental Investigation of the Skin Friction of the Turbulent Boundary Layer on a Flat Plate at Supersonic Speeds," NACA TN 2305, Feb. 1951.

[23]Wardlaw, A. B., Solomon, J. M., and Baltakis, F. P., "Supersonic Inviscid Flowfield Computations of Missile Type Bodies," AIAA Paper 80–0271, Jan. 1980.

[24]Karanian, A. J., Gadbois, S. E., Deblois, R. L., and Lehrach, R. P. C., "Investigation of Air Induction System Technology for an Air Launched Advanced Ramjet Missile, ALARM Test Data Summary Report," UTRC Rept. K910986–15, July 1971.

[25]Kemper, R. G., "As Cast ALVRJ Inlet Test Report," Vought Rept. 2–53390/6R-51310, prepared for the Naval Weapons Center by Vought Corporation System Division, March 1976.

[26]"Refinement on Inlet Design Continues," *Aviation Week and Space Technology*, Vol. 94 No. 6, Feb. 1971, pp. 62–64, special report on Concorde.

[27]Neiner, G. H., Dustin, M. O., and Cole, G. L., "A Throat-Bypass Stability-Bleed System Using Relief Valves to Increase the Transient Stability of a Mixed-Compression Inlet," NASA TP-1083, July 1979.

[28]Mahoney, J. J., "Analytical and Experimental Investigations of Axisymmetric Variable Geometry Inlets for Application to Hypersonic Airbreathing Engines," The Marquardt Co. Rept. 5886, March 1962.

[29]Mahoney, J. J., Alexander, D., McGlone, R. B., and Dunsworth, L. C., "Investigation of a High Performance Hypersonic Inlet," The Marquardt Co. Rept. 5998, Aug. 1963.

[30]Kepler, C. E., Cnossen, J. W., and Demarest, P. E., "Hypersonic Inlet Investigation Including Tests to Mach 8.7 and Theoretical Analyses to Mach 15," United Aircraft Corp., ASD TR-61–137, June 1961.

[31]Wyatt, D. D., "Analysis of Errors Introduced by Several Methods of Weighting Non-Uniform Duct Flows," NACA TN-3400, March 1955.

[32]Stitt, L. E., and Salmi, R. J., "Performance of a Mach 3.0 External-Internal Compression Axisymmetric Inlet at Mach Numbers from 2.0 to 3.5," NASA TMX-145, Jan. 1960.

[33]Mahoney, J. J., and Panek, G. J., "Hypersonic Ramjet Program—Vol. 2: Ramjet Engines for Hypersonic Cruise and Boost Applications," The Marquardt Co. AFAPL TR-65–36, May 1965.

[34]Williams, R. L., "Analytical and Experimental Investigations of a Two-Dimensional Variable Geometry Inlet for Integral Installation on Hypersonic Air Vehicles," The Marquardt Co. Rept. 5902, July 1962.

[35]Heins, A. E., and Hartill, W. R., "Shock Tunnel Testing of Scramjet Inlets," The Marquardt Co. Rept. 5991, May 1963.

[36]Pirrello, C. J., Hardin, R. D., Hechart, M. V., and Brown, K. R., "An Inventory of Aeronautical Ground Research Facilities—Vol. I, Wind Tunnels; Vol. II, Airbreathing Engine Test Facilities; Vol. III, Structural and Environmental Facilities; Vol. IV, Engineering Flight Simulation Facilities," NASA CR-1874, Nov. 1971.

[37]Penaranda, F. E., and Freda, M. S. (editors), "Aeronautical Facilities Catalogue—Vol. I, Wind Tunnels; Vol. II, Airbreathing Propulsion and Flight Simulators," NASA RP-1132, 1985.

INDEX